Machine
to Usin

This *Machinery's Handbook Guide to Using Tables, Formulas, & More in the 32nd Edition* is designed specifically to guide and optimize your use of the latest edition of the *Machinery's Handbook*.

With hundreds of concise discussions, helpful examples, practice exercises, review questions, and answers, the *Guide* addresses a carefully curated selection of the kinds of problems commonly encountered in manufacturing and metalworking.

By following the practical techniques explained and cross-referenced to the *Handbook*, you will become more familiar with the vast range of vital content in the new *32nd Edition*, enhancing your ability to access the information you need and formulate solutions more quickly and easily.

This latest edition of the *Guide* features:

- Added material from the *32nd Edition* and applied revisions throughout.
- Cross references to the *Handbook* and *Digital Edition*–only material.

This companion text is recommended for use alongside the latest and greatest edition of the most popular engineering resource of all time.

USAGE

This edition of the *Machinery's Handbook Guide* should be used in conjunction with the *Machinery's Handbook, 32nd Edition*, which is available in two print sizes, as well as digitally.

This *Guide* is designed to aid in the most efficient use of the *Machinery's Handbook* and to reinforce the extensive information it provides. Hundreds of examples and test questions with answer keys are provided to aid in comprehension of selected topics and guide the use of tables, formulas, and general data in the *Handbook*. This companion volume is designed for use, along with the *32nd Edition*, in general, engineering, and trade schools; home study courses; apprenticeships; and other practical and professional enrichment.

To facilitate usage of this excellent learning and teaching text, *Machinery's Handbook Guide* includes updated page references throughout, both to text and other elements within this edition of the *Guide*, and to the corresponding material in the *Machinery's Handbook, 32nd Edition*, and the *Machinery's Handbook 32 Digital Edition*.

Machinery's Handbook Guide

A Guide to Using Tables, Formulas, & More in the 32nd Edition

By John M. Amiss, Franklin D. Jones,
Henry H. Ryffel, and Christopher J. McCauley

Laura Brengelman, Editor

2024
Industrial Press, Inc.

INDUSTRIAL PRESS, INC.
1 Chestnut Street
South Norwalk, Connecticut 06854 U.S.A.
Phone: 203-956-5593
Toll-Free: 888-528-7852
Email: info@industrialpress.com

Title: *Machinery's Handbook Guide: A Guide to Using Tables, Formulas, & More in the 32nd Edition*
Authors: John H. Amiss, Franklin D. Jones, Henry H. Ryffel, and Christopher J. McCauley
Library of Congress Control Number: 2023949275

ISBN PRINT: 978-0-8311-5032-7
ISBN ePDF: 978-0-8311-9722-3
ISBN ePub: 978-0-8311-9723-0
ISBN: eMobi: 978-0-8311-9823-7

For ISBNs of other digital products and packages including this title, see books.industrialpress.com.

Printed and bound in the United States of America

MACHINERY'S HANDBOOK GUIDE
TO USING THE 32ND EDITION
First Printing

books.industrialpress.com
ebooks.industrialpress.com

TABLE OF CONTENTS

TABLE OF CONTENTS

TABLE OF CONTENTS

TABLE OF CONTENTS

TABLE OF CONTENTS

THE PURPOSE OF THIS BOOK

An engineering handbook is essential equipment for practically all engineers, machine designers, drafters, tool engineers, and skilled mechanics in machine shops and toolrooms. Such a book, with its tables and general data, saves time and labor. To make the best use of any handbook, however, the user must know how to apply the text, tables, formulas, and other data as needed.

One purpose of this *Guide*, which is based on the *Machinery's Handbook*, is to show—through extracted explanations, examples, solutions, and practice questions and answers—typical applications of *Handbook* information, as well as to familiarize engineering students and other users with the *Handbook's* contents. To this end, cross references to material in the *Machinery's Handbook, 32nd Edition,* and in the *Machinery's Handbook 32 Digital Edition* are interspersed throughout this *Guide*. Another objective is to provide test questions and drill work that will enable *Handbook* users, through practice, to both increase their technical knowledge and learn how to obtain needed information quickly and easily.

The *Guide* is available as a paperback, as a standalone eBook, and as part of the *Machinery's Handbook 32 Digital Edition*. In the *Digital Edition* package, the *Guide* includes "live" clickable links to pages, and specific tables, diagrams, and figures in the *Machinery's Handbook* and the *Digital Edition*–only material.

The *Machinery's Handbook*, as with most handbooks, serves as a primary reference, presenting information in condensed form so that a large variety of subjects can be covered in a single volume. Because of this condensed treatment, the practical application of some parts may not always be readily apparent, especially to those who have had little experience in engineering. Therefore, the topical discussions, examples, and questions in this companion volume are not only intended to supplement some of the *Handbook's* content. It is our hope that use of the *Guide* will inform, instruct, and also stimulate interest both in those parts that are referred to regularly and in other interesting and potentially valuable information, even if it is not used that often.

Laura Brengelman
Editor

SECTION 1

DIMENSIONS AND AREAS OF CIRCLES AND SPHERES

Machinery's Handbook pages **78** and **86**

Circumferences of circles are used in calculating speeds of rotating machine parts, including drills, reamers, milling cutters, grinding wheels, gears, and pulleys. These speeds are variously referred to as surface speed, circumferential speed, and peripheral speed; meaning, for each, the distance that a point on the surface or circumference would travel in one minute. This distance usually is expressed as feet per minute. Circumferences are also required in calculating the circular pitch of gears, laying out involute curves, finding the lengths of arcs, and solving many geometrical problems. Letters from the Greek alphabet frequently are used to designate angles, and the Greek letter π (pi) is always used to indicate the ratio between the circumference and the diameter of a circle:

$$\pi = 3.14159265... = \frac{\text{circumference of circle}}{\text{diameter of circle}}$$

For most practical purposes the value of $\pi = 3.1416$ may be used.

Example 1: Find the circumference and area of a circle whose diameter is 8 inches.

On *Handbook* page **78**, the circumference C of a circle is given as $3.1416d$. Therefore, $3.1416 \times 8 = 25.1328$ inches.

On the same page, the area is given as $0.7854d^2$. Therefore, A (area) $= 0.7854 \times 8^2 = 0.7854 \times 64 = 50.2656$ square inches.

Example 2: From page **84** of the *Handbook*, the area of a cylindrical surface equals $S = 3.1416dh$. For a diameter of 8 inches and a height of 10 inches, the area is $3.1416 \times 8 \times 10 = 251.328$ square inches.

Example 3: For the cylinder in **Example 2** but with the area of both ends included, the total area is the sum of the area found in **Example 2** plus two times the area found in **Example 1**.

Thus, $251.328 + 2 \times 50.2656 = 351.8592$ square inches. The same result could have been obtained by using the formula for total area: $A = 3.1416 \times d \times (\frac{1}{2} d + h) = 3.1416 \times 8 \times (\frac{1}{2} \times 8 + 10) = 351.8592$ square inches. Also see the formulas given page **84** of the *Handbook*.

Example 4: If the circumference of a tree is 96 inches, what is its diameter? Since the circumference of a circle $C = 3.1416d$, $96 = 3.1416d$, so $d = 96 \div 3.1416 = 30.558$ inches.

Example 5: The tables starting on page **1074** of the *Handbook* provide values of revolutions per minute required to produce various cutting speeds for workpieces of selected diameters. How are these speeds calculated? Cutting speed in feet per minute is calculated by multiplying the circumference in feet of a workpiece by the rpm of the spindle: cutting speed in fpm = circumference in feet × rpm. By transposing this formula as explained in *Formulas and Their Rearrangement* starting on page **13**,

$$\text{rpm} = \frac{\text{cutting speed, fpm}}{\text{circumference in feet}}$$

For a 3-inch diameter workpiece ($\frac{1}{4}$-foot diameter) and for a cutting speed of 40 fpm, rpm $= 40 \div (3.1416 \times \frac{1}{4}) = 50.92 = 51$ rpm, approximately, which is the same as the value given on page **1074** of the *Handbook*.

Area of Square Inscribed in Circle.—The area of a square inscribed in a circle can be found by drawing a circle and dividing it by two diameters drawn at right angles through the center.

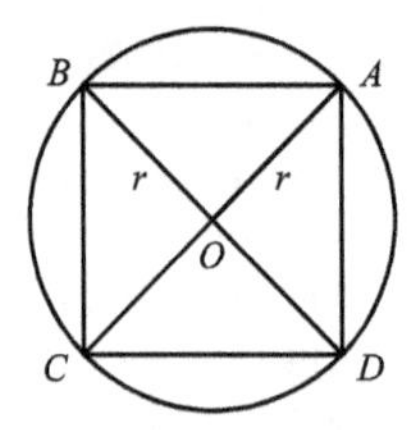

Line $\overline{AB}$ forms one side of the square $ABCD$, and the length of $\overline{AB}$ can be found using the right-triangle formula $a^2 + b^2 = c^2$, *Handbook* page **90**, where lengths a and b are equal to the length of radius r, and c is the length of line $\overline{AB}$.

Therefore,

$$c^2 = a^2 + b^2$$

$$(\overline{AB})^2 = r^2 + r^2 = 2r^2$$

Because the sides of a square are of equal length, the area of the square inscribed in the circle is $\overline{AB} \times \overline{AB} = (\overline{AB})^2 = 2r^2$.

Spheres.—*Handbook* page **86** gives formulas for calculating spherical areas and volumes. Additional formulas are given starting on page **51**.

Example 6: If the diameter of a sphere is $24\frac{5}{8}$ inches, what is the volume, given the formula:

$$\text{Volume} = 0.5236d^3$$

The cube of $24\frac{5}{8}$ = 14,932.369; hence, the volume of this sphere = 0.5236 × 14,932.369 = 7818.5 cubic inches.

Example 7: If the sphere in **Example 6** is hollow and $\frac{1}{4}$ inch thick, what is its weight?

The volume can be obtained using the formula for G_v in the table starting on *Handbook* page **51**, and the density of steel from the table of specific gravity on *Handbook* page **375**.

$$G_v = \frac{4\pi}{3}(R_1^3 - R_2^3) = \frac{\pi}{6}(d_1^3 - d_2^3)$$

$$\text{volume of sphere wall} = \frac{\pi}{6}(24.625^3 - 24.125^3) = 466.65 \text{ in}^3$$

$$466.65 \text{ in}^3 \times \frac{1 \text{ ft}^3}{1728 \text{ in}^3} = 0.27 \text{ ft}^3$$

$$\text{weight of sphere} = 491 \times 0.27 = 132.6 \text{ lb}$$

PRACTICE EXERCISES FOR SECTION 1

(See *Answers to Practice Exercises for Section 1* on page **236**)

1) Find the area and circumference of a circle 10 mm in diameter.

2) On *Handbook* page **1076**, for a 5-mm diameter tool or workpiece rotating at 318 rpm, the corresponding cutting speed is given as 5 meters per minute. Check this value.

3) For a cylinder 100 mm in diameter and 10 mm high, what is the surface area not including the top or bottom?

4) A steel column carrying a load of 10,000 pounds has a diameter of 10 inches. What is the pressure on the floor in pounds per square inch (psi)?

5) What is the ratio of the area of a square of any size to the area of a circle having the same diameter as one side of the square?

6) What is the ratio of the area of a circle to the area of a square inscribed in that circle?

7) The drilling speed for cast iron is assumed to be 70 feet per minute. Find the time required to drill two holes in each of 500 castings if each hole has a diameter of $^3/_4$ inch and is 1 inch deep. Use 0.010-inch feed and allow $^1/_4$ minute per hole for setup.

8) Find the weight of a cast-iron column 10 inches in diameter and 10 feet high. Cast iron weighs 0.26 pound per cubic inch.

9) If machine steel has a tensile strength of 55,000 psi, what should be the diameter of a rod to support 36,000 pounds if the safe working stress is assumed to be one-fifth of the tensile strength?

10) Moving the circumference of a 16-inch automobile flywheel 2 inches moves the camshaft through how many degrees? (The camshaft rotates at one-half the flywheel speed.)

11) What is the area within a circle that surrounds a square inscribed in the circle (shaded area)?

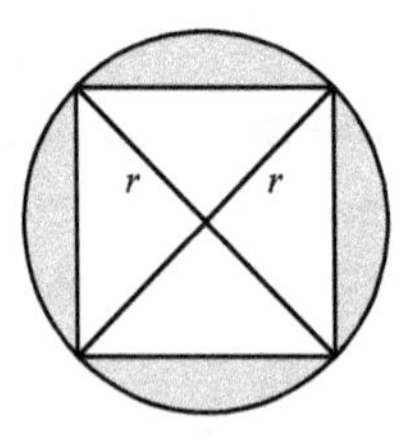

12) A hydraulic cylinder, rated for 2000 psi, is said to have a 5-inch bore and a 2.5-inch piston diameter. If operated at the rated pressure, what forces could be developed on extension and return strokes?

13) Find the number of gallons of oil in a tank 6 feet in diameter and 12 feet long if the tank is in a horizontal position, and the oil measures 2 feet deep.

14) Find the surface areas of the following spheres, the diameters of which are: $1\frac{1}{2}$; $3\frac{3}{8}$; 65; $20\frac{3}{4}$.

15) Find the volume of each sphere in the above exercise.

16) The volume of a sphere is 1,802,725 cubic inches. What are its surface area and diameter?

SECTION 2

CHORDS, SEGMENTS, AND HOLE CIRCLES

Machinery's Handbook pages **50, 79, 85, 702**, and **708–720**

A chord of a circle is the distance along a straight line from one point to any other point on the circumference. A segment of a circle is that part or area between a chord and the arc it intercepts. The lengths of chords and the dimensions and areas of segments are often required in mechanical work.

Lengths of Chords.—The table of chords, *Handbook* **720**, can be applied to a circle of any diameter as explained and illustrated by examples on pages **711** and **720**. The table is given to six decimal places so that it can be used in connection with precision-tool work. Additional related formulas are given on page **702**.

Example 1: A circle has 56 equal divisions, and the chordal distance from one division to the next is 2.156 inches. What is the diameter of the circle?

The chordal length in the table for 56 divisions and a diameter of 1 equals 0.05607; therefore, in this example,

$$2.156 = 0.05607 \times \text{Diameter}$$

$$\text{Diameter} = \frac{2.156}{0.05607} = 38.452 \text{ inches}$$

Example 2: A drill jig is to have eight holes equally spaced around a circle 6 inches in diameter. How can the chordal distance between adjacent holes be determined when the table on *Handbook* page **720** is not available?

One-half the angle between the radial center lines of adjacent holes = 180 ÷ number of holes. If the sine of this angle is multiplied by the diameter of the circle, the product equals the chordal distance. In this example, we have 180 ÷ 8 = 22.5 degrees. The sine of 22.5 degrees from a calculator is 0.38268; hence, the chordal distance = 0.38268 × 6 = 2.296 inches. The result is the same as would be obtained with the table on *Handbook* page **720** because the figures in the column "Length of the Chord" represent the sines of angles equivalent to 180 divided by the different numbers of spaces.

Use of the Table of Segments of Circles—*Handbook* page 79.— This table is of the unit type in that the values all apply to a radius of 1. As explained above the table, the value for any other radius can be obtained by multiplying the figures in the table by the given radius. For areas, the *square* of the given radius is used. Thus, the unit type of table is universal in its application.

Example 3: Find the area of a segment of a circle, the center angle of which is 57 degrees, and the radius $2\frac{1}{2}$ inches.

First locate 57 degrees in the center angle column; opposite this figure in the area column will be found 0.0781. Since the area is required, this number is multiplied by the square of $2\frac{1}{2}$. Thus, $0.0781 \times (2\frac{1}{2})^2 = 0.488$ square inch.

Example 4: A cylindrical oil tank is $4\frac{1}{2}$ feet in diameter and 10 feet long, and is in a horizontal position. When the depth of the oil is 3 feet, 8 inches, what is the number of gallons of oil?

The total capacity of the tank equals $0.7854 \times (4\frac{1}{2})^2 \times 10 = 159$ cubic feet. One US gallon equals 0.1337 cubic foot (see *Handbook* page **2862**); hence, the total capacity of the tank equals $159 \div 0.1337 = 1190$ gallons.

The unfilled area at the top of the tank is a segment having a height of 10 inches or $\frac{10}{27}$ (0.37037) of the tank radius. The nearest decimal equivalent to $\frac{10}{27}$ in Column h of the table starting on page **79** is 0.3707; hence, the number of cubic feet in the segment-shaped space $= (27^2 \times 0.401 \times 120) \div 1728 = 20.3$ cubic feet and $20.3 \div 0.1337 = 152$ gallons. Therefore, when the depth of oil is 3 feet, 8 inches, there are $1190 - 152 = 1038$ gallons. (See also *Handbook* page **68** for additional information on the capacity of cylindrical tanks.)

Example 5: Use the tank from **Example 4** and the table on *Handbook* page **79** to estimate the height of fuel in the tank when the tank contains 150 gallons.

When the tank contains 150 gallons, it is $^{150}/_{1190} \times 100 = 12.6\%$ full. In the table starting on page **79**, locate the value in the $^{A}/_{\pi}$ column closest to 12.6 and find the corresponding value of h. For $^{A}/_{\pi} = 12.4$, $h = .36392$, and the approximate height of fuel in the tank is $h \times r = 0.36392 \times 2.25 \times 12 = 9.83$ inches.

Hole Circle Coordinates.—It is often necessary to manually locate the positions of equally spaced holes around a given hole circle diameter. Pages **712–719** of the *Handbook* provide tables with constants derived using equations on *Handbook* pages **708–710**. These constants can be used to calculate the precise x and y position of each hole.

Type "A" hole circles, as shown in **Fig. 1a** on *Handbook* page **708** and **Fig. 1** below, can be identified by hole number 1 being positioned at the top of the hole circle diameter, with the other holes numbered in a counterclockwise direction. The x and y coordinates of each hole are calculated by multiplying the hole circle diameter by the constants in **Table 1a** on *Handbook* pages **712–713**. The reference point for the x and y coordinates is located at the center of the hole circle diameter.

Example 6: **Fig. 1** shows 7 equally spaced holes on a hole circle diameter of 1.125″ with each hole's x and y coordinates given in the included table at right. The position of hole number 7 is calculated using inch coordinates to 3 decimal places:

$x = 0.39092$ (constant) $\times$ 1.125 (hole circle diameter) $= 0.440$

$y = -0.31174$ (constant) $\times$ 1.125 (hole circle diameter) $= -0.351$

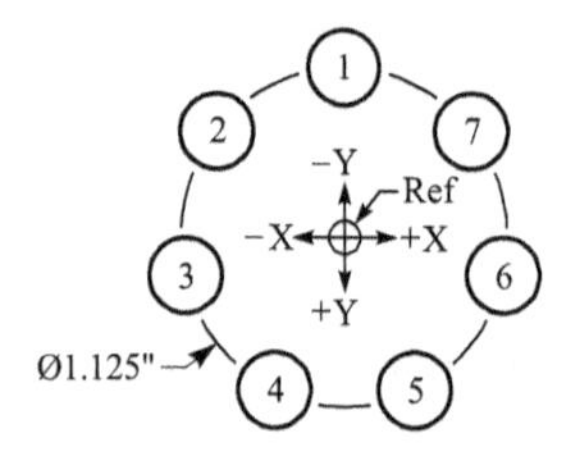

#	x	y
1	0.000	−0.563
2	−0.440	−0.351
3	−0.548	0.125
4	−0.244	0.507
5	0.244	0.507
6	0.548	0.125
7	0.440	−0.351

Fig. 1.

Type "A" hole circles, as shown in **Fig. 1b** on *Handbook* page **708** and in **Fig. 3** below, again can be identified by hole number 1 at the top of the hole circle diameter, with the other holes numbered in a counterclockwise direction. The x and y coordinates of each hole are calculated by multiplying the hole circle diameter by the constants in **Table 1b** on *Handbook* pages **714–715**.

As shown in **Fig. 2**, X0, Y0 is located at the intersection of the vertical x coordinate line and the horizontal y coordinate line, with both lines tangent to the hole circle diameter.

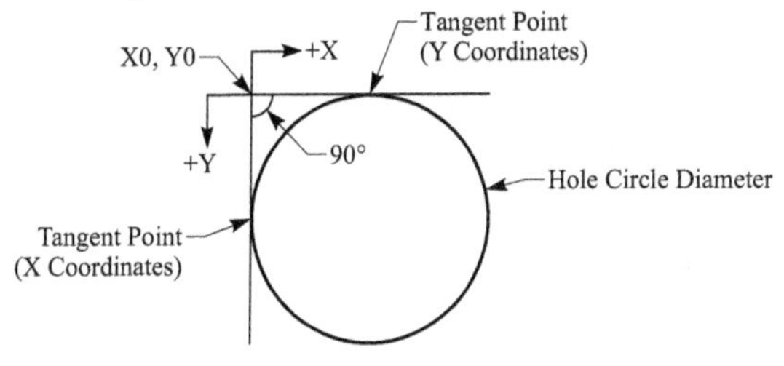

Fig. 2.

Example 7: **Fig. 3** shows 7 equally spaced holes on a hole circle diameter of 1.125″ with each hole's x and y coordinates given in the table at right. The position of hole number 4 is calculated using inch coordinates to 3 decimal places:

$x = 0.28306$ (constant) × 1.125 (hole circle diameter) = 0.318

$y = 0.95048$ (constant) × 1.125 (hole circle diameter) = 1.069

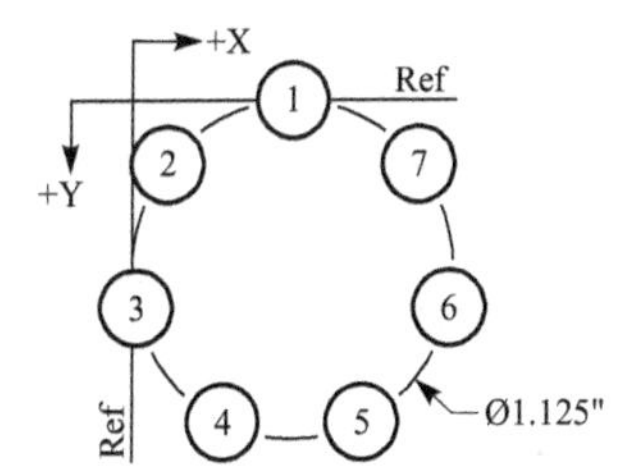

#	x	y
1	0.563	0.000
2	0.123	0.212
3	0.014	0.688
4	0.318	1.069
5	0.807	1.069
6	1.111	0.688
7	1.002	0.212

Fig. 3.

Type "B" hole circles, as shown in **Fig. 2a** on *Handbook* page **709** and in **Fig. 4** below, can be identified by hole number 1 being positioned at the top left of the hole circle diameter, with the other holes numbered in a counterclockwise direction. The x and y coordinates of each hole are calculated by multiplying the hole circle diameter by the constants in **Table 2a** on *Handbook* pages **716–717**. The reference point for the x and y coordinates is located at the center of the hole circle diameter.

Example 8: **Fig. 4** shows 7 equally spaced holes on a hole circle diameter of 1.125″ with each hole's x and y coordinates given in the table at right. The position of hole number 6 is calculated using inch coordinates to 3 decimal places:

$x = 0.48746$ (constant) × 1.125 (hole circle diameter) = 0.548

$y = -0.11126$ (constant) × 1.125 (hole circle diameter) = –0.125

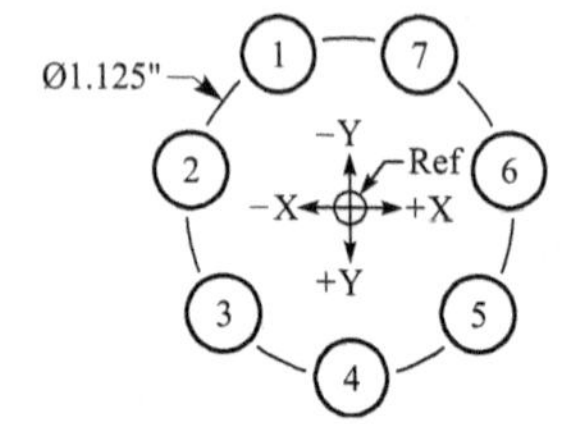

#	x	y
1	–0.244	–0.507
2	–0.548	–0.125
3	–0.440	0.351
4	0.000	0.563
5	0.440	0.351
6	0.548	–0.125
7	0.244	–0.507

Fig. 4.

Type "B" hole circles, as shown in **Fig. 2b** on *Handbook* page **709** and in **Fig. 6** on the next page, again can be identified by hole number 1 at the top left of the hole circle diameter, with the other holes numbered in a counterclockwise direction. The x and y coordinates of each hole are calculated by multiplying the hole circle diameter by the constants in **Table 2b** on *Handbook* pages **718–719**.

As shown in **Fig. 5** on the next page, X0, Y0 is located at the intersection of the vertical x coordinate line and the horizontal y coordinate line, with both lines tangent to the hole circle diameter.

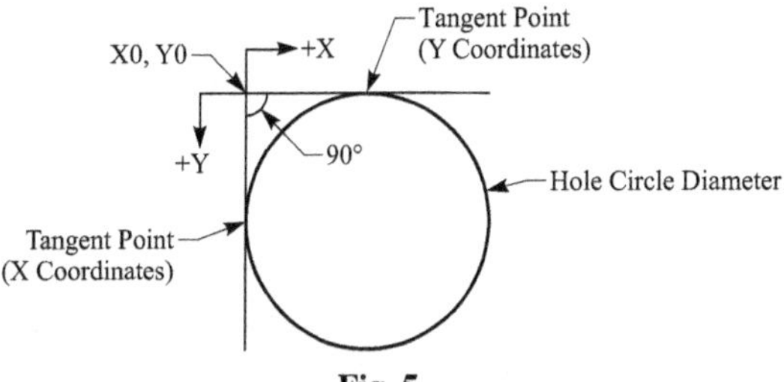

Fig. 5.

Example 9: **Fig. 6** shows 7 equally spaced holes on a hole circle diameter of 1.125″ with each hole's x and y coordinates given in the table at right. The position of hole number 2 is calculated using inch coordinates to 3 decimal places:

$x = 0.01254\text{ (constant)} \times 1.125\text{ (hole circle diameter)} = 0.014$

$y = 0.38874\text{ (constant)} \times 1.125\text{ (hole circle diameter)} = 0.437$

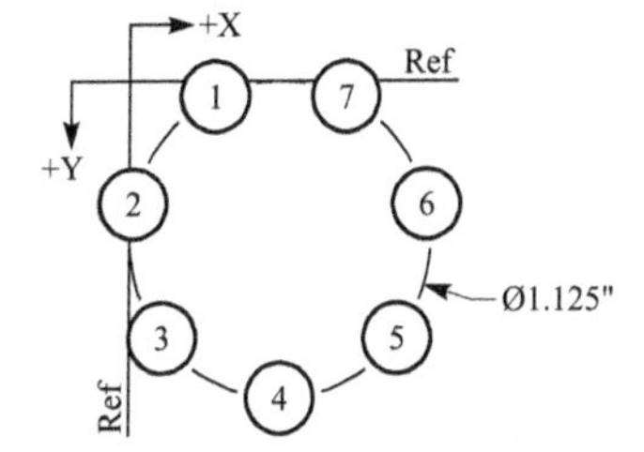

#	x	y
1	0.318	0.056
2	0.014	0.437
3	0.123	0.913
4	0.563	1.125
5	1.002	0.913
6	1.111	0.437
7	0.807	0.056

Fig. 6.

PRACTICE EXERCISES FOR SECTION 2

(See *Answers to Practice Exercises for Section 2* on page **236**)

1) Find the lengths of chords when the number of divisions of a circumference and the radii are as follows: 30 and 4; 14 and $2\frac{1}{2}$; 18 and $3\frac{1}{2}$.

2) Find the chordal distance between the graduations for thousandths on the following dial indicators: (a) Starrett has 100 divisions and $1\frac{3}{8}$-inch dial. (b) Brown & Sharpe has 100 divisions and $1\frac{3}{4}$-inch dial. (c) Ames has 50 divisions and $1\frac{5}{8}$-inch dial.

3) The teeth of gears are evenly spaced on the pitch circumference. In making a drawing of a gear, how wide should the dividers be set to space 28 teeth on a 3-inch-diameter pitch circle?

4) In a drill jig, 8 holes, each $\frac{1}{2}$ inch diameter, were spaced evenly on a 6-inch-diameter circle. To test the accuracy of the jig, plugs were placed in adjacent holes. The distance over the plugs was measured by micrometer. What should be the micrometer reading?

5) In the preceding problem, what should be the distance over plugs placed in alternate holes?

6) What is the length of the arc of contact of a belt over a pulley 2 feet, 3 inches in diameter if the arc of contact is 215 degrees?

7) Find the areas, lengths, and heights of chords of the following segments: (a) radius 2 inches, angle 45 degrees; (b) radius 6 inches, angle 27 degrees.

8) Using **Fig. 1a** Type "A" Circle on page **708** of the *Handbook*, what are the *x* and *y* coordinates to 3 decimal places for hole number 8 in a 9 hole circle, with a hole circle diameter of 5.250 inches?

9) Using **Fig. 1b** Type "A" Circle on page **708** of the *Handbook*, what are the *x* and *y* coordinates to 3 decimal places for hole number 4 in a 5 hole circle, with a hole circle diameter of 3.625 inches?

10) Using **Fig. 2a** Type "B" Circle on page **709** of the *Handbook*, what are the *x* and *y* coordinates to 3 decimal places for hole number 11 in a 17 hole circle, with a hole circle diameter of 4.835 inches?

11) Using **Fig. 2b** Type "B" Circle on page **709** of the *Handbook*, what are the *x* and *y* coordinates to 3 decimal places for hole number 3 in a 5 hole circle, with a hole circle diameter of 2.000 inches?

SECTION 3

FORMULAS AND THEIR REARRANGEMENT

Machinery's Handbook page **21**

A formula may be defined as a mathematical rule expressed by signs and symbols instead of in actual words. In formulas, letters are used to represent numbers or quantities, the term "quantity" being used to designate any number involved in a mathematical process. The use of letters in formulas, in place of the actual numbers, simplifies the solution of problems and makes it possible to condense into small space the information that otherwise would be imparted by long and cumbersome rules. The figures or values for a given problem are inserted in the formula according to the requirements in each specific case. When the values are thus inserted, in place of the letters, the result or answer is obtained by ordinary arithmetical methods. There are two reasons why a formula is preferable to a rule expressed in words. 1) The formula is more concise, it occupies less space, and it is possible to see at a glance the whole meaning of the rule laid down. 2) It is easier to remember a brief formula than a long rule, and it is, therefore, of greater value and convenience.

Example 1: In spur gears, the outside diameter of the gear can be found by adding 2 to the number of teeth and dividing the sum obtained by the diametral pitch of the gear. This rule can be expressed very simply by a formula. Assume that we write D for the outside diameter of the gear, N for the number of teeth, and P for the diametral pitch. Then the formula would be:

$$D = \frac{N+2}{P}$$

This formula reads exactly as the rule given above. It says that the outside diameter (D) of the gear equals 2 added to the number of teeth (N), and this sum is divided by the pitch (P).

If the number of teeth in a gear is 16 and the diametral pitch 6, then simply put these figures in the place of N and P in the formula, and solve for the outside diameter as in ordinary arithmetic.

$$D = \frac{16+2}{6} = \frac{18}{6} = 3 \text{ inches}$$

Example 2: The formula for the horsepower generated by a steam engine is as follows:

$$H = \frac{P \times L \times A \times N}{33{,}000}$$

in which H = indicated horsepower of engine;
P = mean effective pressure on piston in pounds per square inch;
L = length of piston stroke in feet;
A = area of piston in square inches;
N = number of strokes of piston per minute.

Assume that $P = 90$, $L = 2$, $A = 320$, and $N = 110$; what would be the horsepower?

If we insert the given values in the formula, we have:

$$H = \frac{90 \times 2 \times 320 \times 110}{33{,}000} = 192$$

From the examples given, we may formulate the following general rule: *In formulas, each letter stands for a certain dimension or quantity; when using a formula for solving a problem, replace the letters in the formula by the values given for a certain problem, and find the required answer as in ordinary arithmetic.*

Omitting Multiplication Signs in Formulas.—In formulas, the sign for multiplication (×) is often left out between letters the values of which are to be multiplied. Thus AB means $A \times B$, and the formula $H = \frac{P \times L \times A \times N}{33{,}000}$ can also be written $H = \frac{PLAN}{33{,}000}$.

If $A = 3$, and $B = 5$, then: $AB = A \times B = 3 \times 5 = 15$.

It is only the multiplication sign (×) that can be thus left out between the symbols or letters in a formula. All other signs must be indicated the same as in arithmetic. However, the product of two numbers *does* require the multiplication sign, such as 23×4; this also may be written with parentheses, as (23)(4).

As a general rule, the figure in an expression such as "3*A*" is written first and is known as the *coefficient* of *A*.

Rearrangement of Formulas.—A formula can be rearranged or "transposed" to determine the values represented by different letters of the formula. To illustrate by a simple example, the formula for determining the speed (s) of a driven pulley, when its diameter (d) and the diameter (D) and speed (S) of the driving pulley are known, is as follows: $s = (S \times D)/d$. If the speed of the driven pulley is known, and the problem is to find its diameter or the value of d instead of s, this formula can be rearranged or changed. Thus: $d = (S \times D)/s$

Rearranging a formula in this way is governed by four general rules.

Rule 1. An independent term preceded by a plus sign (+) may be transposed to the other side of the equals sign (=) if the plus sign is changed to a minus sign (−).

Rule 2. An independent term preceded by a minus sign may be transposed to the other side of the equals sign if the minus sign is changed to a plus sign.

As an illustration of these rules, if $A = B - C$, then $C = B - A$, and if $A = C + D - B$, then $B = C + D - A$. That the foregoing are correct may be proved by substituting numerical values for the different letters and then transposing them as shown.

Rule 3. A term that multiplies all the other terms on one side of the equals sign may be moved to the other side if it is made to divide all the terms on that side.

As an illustration of this rule, if $A = BCD$, then $A/(BC) = D$ or according to the common arrangement $D = A/(BC)$. Suppose, in the preceding formula, that $B = 10$, $C = 5$, and $D = 3$; then $A = 10 \times 5 \times 3 = 150$ and $150/(10 \times 5) = 3$.

Rule 4. A term that divides all the other terms on one side of the equals sign may be moved to the other side if it is made to multiply all the terms on that side.

To illustrate, if $s = SD/d$, then $sd = SD$, and, according to *Rule 3*, $d = SD/s$. This formula may also be rearranged for determining the values of S and D; thus $ds/D = S$, and $ds/S = D$.

If, in the rearrangement of formulas, minus signs precede quantities, the signs may be changed to obtain positive rather than minus quantities. All the signs on both sides of the equals sign or on both sides of the equation may be changed. For example, if $-2A = -B + C$, then $2A = B - C$. The same result would be obtained by placing all the terms on the opposite side of the equals sign, which involves changing signs. For instance, if $-2A = -B + C$, then $B - C = 2A$.

Fundamental Laws Governing Rearrangement.—After a few fundamental laws that govern any formula or equation are understood, its solution usually is very simple. An equation states that one quantity equals another quantity. So long as both parts of the equation are treated exactly alike, the values remain equal. Thus, in the equation $A = \frac{1}{2}\,ab$, which states that the area A of a triangle equals one-half the product of the base a times the altitude b, each side of the equation would remain equal if we added the same amount: $A + 6 = \frac{1}{2}\,ab + 6$; or we could subtract an equal amount from both sides: $A - 8 = \frac{1}{2}\,ab - 8$; or multiply both parts by the same number: $7A = 7(\frac{1}{2}\,ab)$; or we could divide both parts by the same number, and we would still have a true equation.

One formula for the total area T of a cylinder is:

$$T = 2\pi r^2 + 2\pi rh$$

where: r = radius and h = height of the cylinder. Suppose we want to solve this equation for h. Transposing the part that does not contain h to the other side by changing its sign, we get: $2\pi rh = T - 2\pi r^2$. To obtain h, we can divide both sides of the equation by any quantity that will leave h on the left-hand side; thus:

$$\frac{2\pi rh}{2\pi r} = \frac{T - 2\pi r^2}{2\pi r}$$

It is clear that, in the left side of the equation, the $2\pi r$ will cancel out, leaving: $h = (T - 2\pi r^2)/(2\pi r)$. The expression $2\pi r$ on the right side cannot be cancelled because it is not an independent factor, since the numerator equals the difference between T and $2\pi r^2$.

Example 3: Rearrange the formula for a trapezoid (*Handbook* page **72**) to obtain h.

$$A = \frac{(a+b)h}{2}$$

$2A = (a+b)h$ (multiply both sides of the equation by 2)

$(a+b)h = 2A$ (transpose both sides so as to get the multiple of h on the left side)

$\frac{(a+b)h}{a+b} = \frac{2A}{a+b}$ (divide both sides by $a+b$)

$h = \frac{2A}{a+b}$ (cancel $a+b$ from the left side)

Example 4: The formula for determining the radius of a sphere (*Handbook* page **86**) is as follows:

$$r = \sqrt[3]{\frac{3V}{4\pi}}$$

Rearrange to obtain a formula for finding the volume V.

$r^3 = \frac{3V}{4\pi}$ (cube each side)

$4\pi r^3 = 3V$ (multiply each side by 4π)

$3V = 4\pi r^3$ (transpose both sides)

$\frac{3V}{3} = \frac{4\pi r^3}{3}$ (divide each side by 3)

$V = \frac{4\pi r^3}{3}$ (cancel 3 from the left side)

The procedure has been shown in detail to indicate the underlying principles involved. Rearrangement could be simplified somewhat by direct application of the rules previously given.

To illustrate:

$$r^3 = \frac{3V}{4\pi} \quad \text{(cube each side)}$$

$$4\pi r^3 = 3V \quad \text{(applying } \textit{Rule 4} \text{ move } 4\pi \text{ to the left side)}$$

$$\frac{4\pi r^3}{3} = V \quad \text{(move 3 to the left side—} \textit{Rule 3}\text{)}$$

This final equation would, of course, be reversed to locate V on the left side of the equals sign, as this is the usual position for whatever letter represents the quantity or value to be determined.

Example 5: Determine the diameter of cylinder and length of stroke of a steam engine to deliver 150 horsepower. The mean effective steam pressure is 75 pounds, and the number of strokes per minute is 120. The length of the stroke is to be 1.4 times the diameter of the cylinder.

First, insert the known values into the horsepower formula (**Example 2**, page **14**):

$$150 = \frac{75 \times L \times A \times 120}{33{,}000} = \frac{3 \times L \times A}{11}$$

The last expression is found by cancellation.

Assume now that the diameter of the cylinder in inches equals D. Then, $L = 1.4D/12 = 0.117D$ according to the requirements in the problem; the divisor 12 is introduced to change the inches to feet (L being in feet in the horsepower formula). The area $A = D^2 \times 0.7854$. If we insert these values in the last expression in our formula, we have:

$$150 = \frac{3 \times 0.117D \times 0.7854D^2}{11} = \frac{0.2757D^3}{11}$$

$$0.2757D^3 = 150 \times 11 = 1650$$

$$D^3 = \frac{1650}{0.2757} \qquad D = \sqrt[3]{\frac{1650}{0.2757}} = \sqrt[3]{5984.8} = 18.15$$

The diameter of the cylinder should be about 18¼ inches, and the length of the stroke $18.15 \times 1.4 = 25.41$, or about 25½ inches.

Solving Equations or Formulas by Trial.—One of the equations used for spiral gear calculations, when the shafts are at right angles, the ratios are unequal, and the center distance must be exact, is as follows:

$$R \sec\alpha + \csc\alpha = \frac{2CP_n}{n}$$

In this equation

R = ratio of number of teeth in large gear to number in small gear

C = exact center distance

P_n = normal diametral pitch

n = number of teeth in small gear

The exact spiral angle α of the large gear is found by trial using the equation just given.

Equations of this form are solved by trial by selecting an angle assumed to be approximately correct and inserting the secant and cosecant of this angle in the equation, adding the values thus obtained, and comparing the sum with the known value to the right of the equals sign in the equation. An example will show this more clearly. By using the problem given in the *Example* near the top of *Handbook* page **2282** to illustrate, $R = 3$; $C = 10$; $P_n = 8$; $n = 28$.

Hence, the whole expression $\frac{2CP_n}{n} = \frac{2 \times 10 \times 8}{28} = 5.714$ from which it follows that:

$$R \sec\alpha + \csc\alpha = 5.714$$

In the problem given, the spiral angle required is 45 degrees. The spiral gears, however, would not meet all the conditions given in the problem if the angle could not be slightly modified. To determine whether the angle should be greater or smaller than 45 degrees, insert the values of the secant and cosecant of 45 degrees in the formula. The secant of 45 degrees is 1.4142, and the cosecant is 1.4142. Then,

$$3 \times 1.4142 + 1.4142 = 5.6568$$

The value 5.6568 is too small, as it is less than 5.714, which is the required value. Hence, try 46 degrees. The secant of 46 degrees is 1.4395, and the cosecant, 1.3902.

Then,

$$3 \times 1.4395 + 1.3902 = 5.7087$$

Obviously, an angle of 46 degrees is also too small. Proceed, therefore, to try an angle of 46 degrees, 30 minutes. This angle will be found too great. Similarly 46 degrees, 15 minutes, if tried, will be found too great, and by repeated trials it will finally be found that an angle of 46 degrees, 6 minutes, the secant of which is 1.4422, and the cosecant, 1.3878, meets the requirements. Then,

$$3 \times 1.4422 + 1.3878 = 5.7144$$

which is as close to the required value as necessary.

In general, when an equation must be solved by the trial-and-error method, all the known quantities may be written on the right side of the equal sign and all the unknown quantities on the left side. A value is assumed for the unknown quantity. This value is substituted in the equation, and all the values thus obtained on the left side are added. In general, if the result is greater than the values on the right side, the assumed value of the unknown quantity is too great. If the result obtained is smaller than the sum of the known values, the assumed value for the unknown quantity is too small. By thus adjusting the value of the unknown quantity until the left side of the equation with the assumed value of the unknown quantity will just equal the known quantities on the right side of the equal sign, the correct value of the unknown quantity may be determined.

Derivation of Formulas.—Most formulas in engineering handbooks are given without showing how they have been derived or originated, because engineers and designers usually want only the final results; moreover, such derivations would require considerable additional space, and they belong in textbooks rather than in handbooks, which are primarily works of reference. Although *Machinery's Handbook* contains thousands of formulas derived from published Standards and other special formulas, it is apparent that no handbook can include every kind of formula, because a great many formulas apply only to local designing or manufacturing problems. Such special formulas are derived by engineers and designers for their own use. The exact methods of deriving formulas are based upon mathematical principles as they are related to the particular factors that apply. A few examples will be given to show how several different types of special formulas have been derived.

Example 6: The problem is to deduce the general formula for finding the point of intersection of two tapers with reference to measured diameters on those tapers. In **Fig. 1**,

L = distance between the two measured diameters, D and d

X = the required distance from one measured diameter to the intersection of tapers

a = angle of long taper as measured from center line

a_1 = angle of short taper as measured from center line

Then,

$$E = \frac{D-d}{2} = Z + Y$$

$$Z = (L-X)\tan a_1$$

$$Y = X \tan a$$

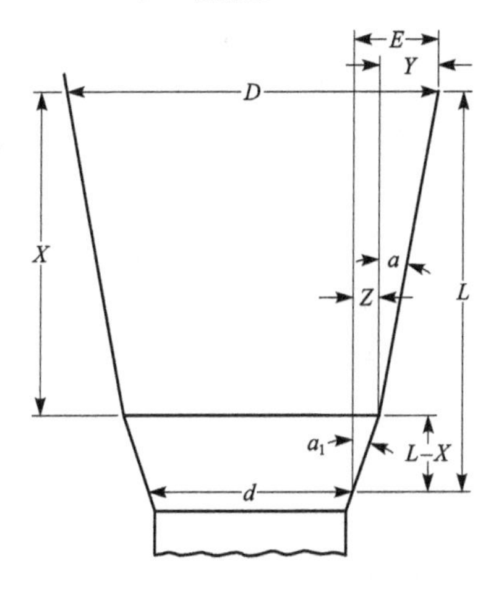

Fig. 1. To Find Dimension *X* from a Given Diameter *D* to the Intersection of Two Conical Surfaces

Therefore:

$$\frac{D-d}{2} = (L-X)\tan a_1 + X\tan a$$

and

$$D - d = 2\tan a_1 (L-X) + 2X\tan a \qquad (1)$$

But

$$2\tan a_1 = T_1 \qquad \text{and} \qquad 2\tan a = T$$

in which T and T_1 represent the long and short tapers per inch, respectively.

Therefore, from **Equation (1)**,

$$D - d = T_1 (L - X) + TX$$

$$D - d = T_1 L - T_1 X + TX$$

$$X(T_1 - T) = T_1 L - (D - d)$$

$$X = \frac{T_1 L - (D - d)}{T_1 - T}$$

Example 7: A flywheel is 16 feet in diameter (outside measurement), and the center of its shaft is 3 feet above the floor. Derive a formula for determining how long the hole in the floor must be to permit the flywheel to turn.

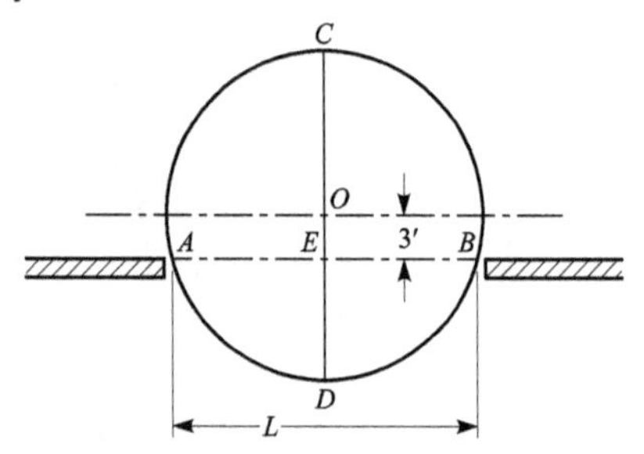

Fig. 2. To Find Length of Hole in Floor for Flywheel

The conditions are as represented in **Fig. 2**. The line AB is the floor level and is a chord of the arc ABD; it is parallel to the horizontal diameter through the center O. CD is the vertical diameter and is perpendicular to AB. It is shown in geometry that the diameter CD bisects the chord AB at the point of intersection E.

One of the most useful theorems of geometry is that when a diameter bisects a chord, the product of the two parts of the diameter is equal to the square of one half the chord; in other words,

$(AE)^2 = ED \times EC$. If AB is represented by L and OE by a, $ED = r - a$ and $EC = r + a$, in which r = the radius OC; hence,

$$\left(\frac{L}{2}\right)^2 = (r-a)(r+a) = r^2 - a^2$$

$$\frac{L}{2} = \sqrt{r^2 - a^2} \text{ and } L = 2\sqrt{r^2 - a^2}$$

By substituting the values given,

$$L = 2\sqrt{8^2 - 3^2} = 14.8324 \text{ feet} = 14 \text{ feet, } 10 \text{ inches}$$

The length of the hole, therefore, should be at least 15 feet, to allow sufficient clearance.

Empirical Formulas.—Many formulas used in engineering calculations cannot be established fully by mathematical derivation but must be based upon actual tests instead of relying upon mere theories or assumptions that might introduce excessive errors. These formulas are known as *empirical formulas*. Usually such a formula contains a constant (or constants) that represents the result of the tests; consequently, the value obtained by the formula is consistent with these tests or with actual practice.

A simple example of an empirical formula is found in the following formula for the breaking load of chain in pounds, given on *Machinery's Handbook 32 Digital Edition* page **3793**:

$$W = 54{,}000D^2$$

This particular formula contains the constant 54,000, which was established by tests, and the formula is used to obtain the breaking load of wrought-iron crane chains to which a factor of safety of 3, 4, or 5 is then applied to obtain the working load.

Handbook page **289** contains an example of an empirical formula based upon experiments made with power-transmitting shafts. This formula gives the diameter of shaft required to prevent excessive twisting during transmission of power.

Parentheses.—Two important rules related to the use of parentheses are based upon the principles of positive and negative numbers:

1) If a parenthesis is preceded by a + sign, it may be removed, and the terms within the parentheses will retain their signs.

$$a + (b - c) = a + b - c$$

2) If a parenthesis is preceded by a – sign, it may be removed, but only if the signs preceding each of the terms inside of the parentheses are changed (+ changed to –, and – to +). Multiplication and division signs are not affected.

$$a - (b - c) = a - b + c$$

$$a - (-b + c) = a + b - c$$

Knowledge of algebra is not necessary to make successful use of formulas of the general type such as are found in engineering handbooks; it is only necessary to understand thoroughly the use of letters or symbols in place of numbers, and to be well versed in the methods, rules, and processes of ordinary arithmetic. Knowledge of algebra becomes necessary only where a general rule or formula that gives the answer to a problem directly is not available. In other words, algebra is useful in *developing* or originating a general rule or formula, but the formula can be *used* without recourse to algebraic processes.

Constants.—A constant is a value that does not change or is not variable. Constants at one stage of a mathematical investigation may be variables at another stage, but an *absolute constant* has the same value under all circumstances. The ratio of the circumference to the diameter of a circle, or 3.1416, is a simple example of an absolute constant. In a common formula used for determining the indicated horsepower of a reciprocating steam engine, the product of the mean effective pressure in pounds per square inch, the length of the stroke in feet, the area of the piston in square inches, and the number of piston strokes per minute is divided by the constant 33,000, which represents the number of foot-pounds of work per minute equivalent to 1 horsepower. Constants occur in many mathematical formulas.

Mathematical Signs and Abbreviations.—Every division of mathematics has its traditions, customs, and signs that are frequently of ancient origin. Hence, we encounter Greek letters in many problems (see *Handbook* page **2836**) where it would seem that English letters would do as well or better. Many of the signs and abbreviations on *Handbook* pages **2836** to **2838** will be used frequently. They should, therefore, be understood.

Conversion Tables.—It may sometimes be necessary to convert English units of measurement into metric units and vice versa. The tables provided at the back of the *Handbook* will be found useful in this connection. A table of metric conversion factors is also provided in this book starting on page **268**.

PRACTICE EXERCISES FOR SECTION 3

(See *Answers to Practice Exercises for Section 3* on page **237**)

1) An approximate formula for determining the horsepower H of automobile engines is: $H = D^2SN/3$, where D = diameter of bore, inches; S = length of stroke, inches; and N = number of cylinders. Find the horsepower of the following automobile engine: a) bore, $3\frac{1}{2}$ inches; stroke, $4\frac{1}{4}$ inches; cylinders, 6. b) By using the reciprocal of 3, how could this formula be stated?

2) Using the right-angle triangle formula: $C = \sqrt{a^2 + b^2}$, where a = one leg of the triangle, b = the other leg, and C = the hypotenuse, find the hypotenuse of a right triangle whose legs are 16 inches and 63 inches.

3) The formula for finding the blank diameter of a cylindrical shell is: $D = \sqrt{d \times (d + 4h)}$, where D = blank diameter, d = diameter of the shell, and h = height of the shell. Find the diameter of the blank to form a cylindrical shell of 3 inches in diameter and 2 inches high.

4) If D = diagonal of a cube, d = diagonal of face of a cube, s = side of a cube, and V = volume of a cube, then $d = \sqrt{2D^2/3}$; $s = \sqrt{D^2/3}$; and $V = s^3$. Find the side, volume of a cube, and diagonal of the face of a cube if the diagonal of the cube is 10.

5) The area of an equilateral triangle equals one fourth of the square of the side times the square root of 3, or

$$A = (S^2/4)\sqrt{3} = 0.43301S^2$$

Find the area of an equilateral triangle whose side is 14.5 inches.

6) The formula for the volume of a sphere is: $4\pi r^3/3$ or $\pi d^3/6$. What constants may be used in place of $4\pi/3$ and $\pi/6$?

7) The formula for the volume of a solid ring is $2\pi^2 Rr^2$, where r = radius of cross section and R = radius from the center of the ring to the center of the cross section. Find the volume of a solid ring made from 2-inch round stock if the mean diameter of the ring is 6 inches.

8) Explain these signs: $\pm, >, <, \sin^{-1}a, \tan, \angle, \sqrt[4]{\ }, \log, \theta, \beta, ::$

9) The area A of a trapezoid (see *Handbook* page **72**) is found by the formula:

$$A = \frac{(a+b)h}{2}$$

Transpose the formula for determining width a.

10) $R = \sqrt{r^2 + s^2/4}$; solve for r.

11) $P = 3.1416\sqrt{2(a^2 + b^2)}$; solve for a.

12) $\cos A = \sqrt{1 - \sin^2 A}$; solve for $\sin A$.

13) $a/\sin A = b/\sin B$; solve for $a, b, \sin A, \sin B$.

SECTION 4

SPREADSHEET CALCULATIONS

Spreadsheet computer programs or spreadsheets are versatile, powerful tools for doing repetitive or complicated algebraic calculations. They are used in diverse technological fields including manufacturing, design, and finance. Spreadsheets blend the power of high-level computer languages with the simplicity of hand calculators. They are ideal for doing "what-if" calculations, such as changing a problem's parameters and comparing the new result to the initial answer. The visual nature of spreadsheets allows the user to grasp quickly and simultaneously the interaction of many variables in a given problem.

Generally only 5 to 10 percent of a spreadsheet program functionality needs to be understood to begin doing productive spreadsheet calculations. Since the underlying concepts of all spreadsheets are the same, it is easy to transfer this basic understanding from one spreadsheet program to another with a minimal learning curve. Only a small percentage of the actual spreadsheet commands will be covered in this section, but understanding these core concepts will allow the reader to do productive work immediately.

There are many varieties of spreadsheet programs. It is impossible to cover all these spreadsheet programs individually in this brief overview. The formulas listed below are for conceptual understanding and may not work when plugged directly into a particular program. The user should consult the spreadsheet's manual or built-in help system for examples. Generally for any given topic a spreadsheet's help system will list a properly constructed example of what the user is trying to do. The reader can use this as a guide and template to get started.

Basic Spreadsheet Concepts.—To begin using spreadsheets, several key spreadsheet concepts must be understood.

Cell Content: The basic calculating unit of all spreadsheets are cells. Cells may either contain formulas, which are discussed further on; or numbers, words, dates, percentages, and currency. A cell normally has to be formatted using the spreadsheet's cell format commands to display its contents correctly. The formatting usually does not affect the internal representation of the cell, e.g. the actual value of the number. For example, a cell formatted as a *percentage* such as "12%" would actually contain a value of "0.12" in the cell. If the cell were left unformatted "0.12" would be displayed. A cell formatted for currency would display "3.4" as "$3.40,"

Number	*Currency*	*Text*	*Percentage*
12.7854	$12.05	Feed Rate	12% *or* 0.12

Cells containing numbers may be formatted to display an arbitrary level of precision. Again the displayed precision has no affect on actual calculations. For example, the contents of a particular cell containing "3.1415" could be formatted to display "3.141" or "3.14" or "3". Regardless of what is displayed "3.1415" will be used internally by the program for all calculations that refer to that cell.

Formatting cells, while not absolutely necessary, is usually a good idea for several reasons. Formatted cells help others understand your spreadsheet. "12%" is easily identifiable as an interest rate, "0.12" is not. Formatting can also help to avoid input mistakes in large spreadsheets, such as accidently placing an interest rate percentage in a payment currency-formatted cell. The interest rate will be displayed as "$0.12," immediately telling the user something is wrong. For quick "back-of-the-envelope" calculations, formatting can be dispensed with to save time.

Cell Address: In addition to content, cells also have addresses. A cell address is created by combining the column and row names of that cell. In the spreadsheet in **Table 1a**, *Parts* would have an address of *A1*, *Machine 2* would be *C1*, and "13.76" would be *B3*. Spreadsheets use these cell addresses to combine and manipulate the cell contents using formulas.

Table 1a. Machine Cost Spreadsheet (Display)

	A	B	C	D
1	Parts	Machine 1	Machine 2	Total
2	Motor	12.89	$18.76	$31.65
3	Controls	13.76	$19.56	$33.32
4	Chassis	15	$21.87	$36.87
5	Rebate	–7.5	–$10.00	–$17.50
6	Total	34.15	$50.19	$84.34

Formulas: Instead of containing values, a cell may have a formula assigned to it. Spreadsheets use these formulas to manipulate, combine, and chain cells mathematically. The specific format or syntax for properly constructing a formula varies from spreadsheet to spreadsheet. The two most common formula construction techniques are illustrated using the spreadsheet in **Table 1b**.

Table 1b. Machine Cost Spreadsheet (Formulas)

	A	B	C	D
1	Parts	Machine 1	Machine 2	Total
2	Motor	12.89[a]	$18.76	=+B2+C2[b] =$31.65
3	Controls	13.76[a]	$19.56	=Sum(B3:C3)[b] =$33.32
4	Chassis	15[a]	$21.87	=Sum(B4:C4)[b] =$36.87
5	Rebate	–7.5[a]	–$10.00	=Sum (B5:C5)[b] =–$17.50
6	Total	=+B2+B3+B4 +B5[b] =Sum(B2:B5) =34.15[a]	=Sum(C2:C5)[b] =$50.19	=Sum(D2:D5)[b,c] =Sum(B6:C6)[d] =$84.34

[a] This cell is unformatted. This does not change the value of the intermediate calculations or final results.

[b] Cells cannot contain more than one value or formula. The double values and formulas listed in this cell are for illustration only and would not be allowed in a working spreadsheet.

[c] Sum of the machine *Parts*.

[d] Sum of *Machine 1* and *Machine 2*.

Cell by Cell: Each cell is added, subtracted, multiplied, or divided individually. For example in **Table 1b**, the total cost of *Machine 1* would be the values of each individual part cost in column *B* added vertically in cell *B6*.

B6=+B2+B3+B4+B5=$34.15

Sum Function: For long columns or rows of cells, individual cell addition becomes cumbersome. Built-in functions simplify manipulating multiple cells by applying a specific function, like addition, over a range of cells. All spreadsheets have a summation or *Sum* function that adds all the cells that are called out in the function's address range. The *Sum* function adds cells horizontally or vertically. Again in **Table 1b**, the total cost of *Machine 1* using the *Sum* function would be:

B6=Sum(B2:B5)=$34.15

Either method yields the same result and may be used interchangeably. The cell-by-cell method must be used for cells that are not aligned horizontally or vertically. The compact *Sum* method is useful for long chains or ranges of cells. Spreadsheets contain many, many built-in functions that work with math, text strings, dates, etc.

Adding Formulas: Cells containing formulas can themselves be combined, i.e. formulas containing formulas. In **Table 1b**, the total of the *motor* parts (row 2) for *Machine 1* and *Machine 2*, is calculated by the formula in cell *D2*, the total of the *control* parts *D3*, the total of all *chassis* parts *D4*, and the total of the *rebates* in *D5*. These formulas are summed together vertically in the first formula in cell *D6* to get the total cost of all the parts, in this case $84.34. Note that a spreadsheet cell may only contain one formula or value. The multiple formulas in *D6* are for illustration only.

Alternatively, the cost of *Machine 1*, *B6* and *Machine 2*, *C6* could be added together horizontally to get the cost of all the machines, which, in this case, equals the cost of all parts $84.34. This illustrates that it is possible to set up a spreadsheet to find a solution in more than one way. In this case the total cost of all machines was calculated by adding the parts' subtotals or the individual machines' subtotals.

Positive and Negative: Spreadsheets usually display negative numbers with a minus sign (–) in front of them. Sometimes a negative cell number may be formatted to display parentheses around a number instead of a minus sign. For example, –12.874 would be equivalent to (12.874). As with general formatting, this has no effect on the actual cell value.

It is extremely important to treat positive and negative cell values consistently. For example, cell values representing a loan amount of $22,000 and a payment of $500 might be entered as +$22,000 and –$500 if you are receiving a loan or –$22,000 and +$500 if you are loaning the money to someone. Switching one of the signs will create an error in the spreadsheet.

Generally it doesn't matter how positive and negative numbers are assigned, so long as the user is consistent throughout the spreadsheet and the people using the spreadsheet understand the positive-negative frame of reference. Failure to be consistent will lead to errors in your results.

Basic Mathematical Operators: Spreadsheets generally use the following conventions for basic mathematical operators. These operators may be applied to cell values or cell formulas.

Basic Spreadsheet Mathematical Operators

Function	Operator	Function	Operator
Add	+	Divide	/
Subtract	–	Square	^2
Multiply	*	Square Root	^.5
Grouping	((5+B2)/A2) – (6*((9+16)^0.5))		

Consult the spreadsheet's help system to properly construct other mathematical operations, such as sine, cosine, tangent, logarithms, etc.

Built-In Functions: As previously mentioned, spreadsheets contain many built-in functions to aid the user in setting up equations. For example, most spreadsheets have built-in interest functions sometimes referred to as Time Value of Money or *TVM* equations. Generally the names of the variables in the built-in equations do not always exactly match the generally accepted mathematical names used in particular field, such as economics.

To illustrate this point, let's compare the *TVM* terms found in *Variables* on *Handbook* page **138** to the variable names found in a spreadsheet's future value (*FV*) built-in function. Then we'll redo the *Compound Interest* problem found on *Handbook* page **139**.

Example 1, Compound Interest: At 10 percent interest compounded annually for 3 years, a principal amount *P* of \$1000 becomes a sum $F = 1000(1 + 10/100)^3 = \$1,331.93$.

To solve this problem using a spreadsheet, use the future value, *FV* built-in equation *FV*(*Rate*, *Nper*, *Pmt*, *Pv*) where

FV = *F* or the future value of the amount owed or received.

Rate = *I* or nominal annual interest rate per period. In this yearly case, divide by 1; for monthly payments, divide by 12.

Nper = *n* or number of interest periods—in this case, 3. If the interest were compounded monthly, then *Nper* = 3 years × 12 periods /yr. = 36 periods

Pmt = *R* or the payments made or received. For a compound interest loan, *Pmt* =\$0.00

PV = *P* or principal amount lent or borrowed.

To solve, plug in the appropriate values. Again note that leaving column *B* unformatted or formatting column *C* makes no difference for the final answer, but does make it easier to understand the spreadsheet values.

Table 2. Compound Interest Calculations Spreadsheet

	A	B	C	D
1		Value	Value	
2	*Rate*	.1[a]	10%[b]	
3	*Nper*	3[a]	3[b]	
4	*Pmt*	0[a]	\$0.00[b]	
5	*PV*	−1000[a,c]	−\$1,000.00[b,c]	
6	*FV*	= *FV*(B2,B3,B4,B5) = 1,331.93[a]	= \$1,331.93[b]	

[a] Unformatted cell.
[b] Formatted cell.
[c] This number is negative because you are loaning the money out to collect interest.

Advanced Spreadsheet Concepts.—One of the great strengths of spreadsheets is their ability to quickly and easily do what-if calculations. The two key concepts required to do this are cell content and formula "copying and pasting" and "relative and absolute" cell addressing.

Copying and Pasting: Spreadsheets allow cells to be moved, or copied and pasted into new locations. Since a chain of cells can represent a complete problem and solution, copying these chains and pasting them repeatedly into adjacent areas allows several experimental what-if scenarios to be set up. It is then easy to vary the initial conditions of the problem and compare the results side by side. This is illustrated in the following example.

Example 2, What-If Compound Interest Comparison: Referring back to the compound interest problem in **Example 1**, compare the effects of different interest rates from three banks using the same loan amount and loan period. The banks offer a 10%, 11%, and 12% rate. In the spreadsheet, enter 10%, 11%, and 12% into *B2*, *C2*, and *D2*, respectively. Instead of typing in the initial amounts and formulas for the other values for other banks, type them in once in *B3*, *B4*, *B5*, and *B6*. Copy these cells one column over, into column *C* and column *D*. The spreadsheet will immediately solve all three interest rate solutions.

Table 3. Interest Calculations Spreadsheet Using Relative Addressing

	A	B	C	D	E
1	Term	Bank A	Bank B	Bank C	
2	*Rate*	10%	11%	12%	4 cells above "relative" to **E5**
3	*Nper*	3	3	3	3
4	*Pmt*	$0.00	$0.00	$0.00	2
5	*PV*	–$1,000	–$1,000	–$1,000	1
6	*FV*	=*FV*(**B2**,**B3**, **B4**,**B5**) =$1,331.93	=*FV*(**C2**,**C3**, **C4**,**C5**) =$1,367.63	=*FV*(**D2**,**D3**, **D4**,**D5**) =$1,404.93	Cell **E5**

Relative versus Absolute Address: In row 6 of **Table 3**, notice how the *FV* function cell addresses changed as they were copied from column *B* and pasted into the columns *C* and *D*. The formula cell addresses were changed from **B** to **C** in column *C* and **B** to **D** in column *D*. This is known as *relative addressing*. Instead of the formulas pointing to the original or *absolute* locations in the *B* column they were changed by the spreadsheet program as they were pasted to match a cell location with the same relative distance and direction as the original cell. To clarify, in column *E*, the cell *E2* is 4 cells up relative to *E5*. Relative addressing while pasting allows spreadsheets users to easily copy and paste multiple copies of a series of calculations. This easy what-if functionality is a cornerstone of spreadsheet usefulness.

Absolute Addressing: For large complicated spreadsheets the user may want to examine several what-if conditions while varying one basic parameter. For this type of problem it is useful to use *absolute addressing*. There are several formats for creating absolute addresses. Some spreadsheets require a "$" be placed in front of each address. The relative address "*B2*" would become an absolute address when entered as "*B2*". When a formula with an absolute address is copied and pasted the copied formula refers to the same cell as the original. The power of this is best illustrated by an example. Formulas can use a combination of relative and absolute addressing, as in $B2 or B$2, where the absolute portion of the cell address is preceded by the $ symbol.

Example 3, Absolute and Relative Addressing: Suppose that in **Example 1** we wanted to find the future value of $1,000, $1,500 and $2,000 for 10% and 11% interest rates. Using the previous example as a starting point we enter values for *Rate*, *Nper*, *Pmt*, and *Pv*. We also enter the function *FV* into cell *B6*. This time we enter the absolute address *B2* for the *Rate* variable. Now when we copy cell *B6* into *C6* and *D6*, the Rate variable continues to point to cell B2 (absolute addresses) while the other variables *Nper*, *Pmt*, and *Pv* point to locations in columns *C* and *D* (relative addresses).

From the **Table 4a** we find the future value for different starting amounts for a 10% rate. We change cell *B2* from 10% to 11%, and the spreadsheet updates all the loan calculations based on the new interest rate.

Table 4a. 10% Interest Rate Calculations Spreadsheet Using Absolute Addressing

	A	B	C	D
1	Term	Loan Amount A	Loan Amount B	Loan Amount C
2	*Rate*	10%		
3	*Nper*	5	4	3
4	*Pmt*	$0.00	$0.00	$0.00
5	*PV*	–$1,000	–$1,500	–$2,000
6	*FV*	=*FV*(**B2**,B3, B4,B5) =$1,610.51	=*FV*(**B2**,C3, C4,C5) =$2,196.15	=*FV*(**B2**,D3, D4,D5) =$2,662.00

These new values are displayed in **Table 4b**. All we had to do was change one cell to try a new what-if. By combining relative and absolute addresses, we were able to compare the effects of three different loan amounts using two interest rates by changing one cell value.

Table 4b. 11% Interest Rate Calculations Spreadsheet Using Absolute Addressing

	A	B	C	D
1	Term	Loan Amount A	Loan Amount B	Loan Amount C
2	*Rate*	11%		
3	*Nper*	5	4	3
4	*Pmt*	$0.00	$0.00	$0.00
5	*PV*	–$1,000	–$1,500	–$2,000
6	*FV*	=*FV*(**B2**,B3, B4,B5) =$1,685.06	=*FV*(**B2**,C3, C4,C5) =$2,277.11	=*FV*(**B2**,D3, D4,D5) =$2,735.26

Other Capabilities: In addition to mathematical manipulations, most spreadsheets can create graphs, work with dates and text strings, link results to other spreadsheets, and create conditional programming algorithms, to name a few advanced capabilities. While these features may be useful in some cases, many real-world problems can be solved with spreadsheets using simple operators and concepts.

PRACTICE EXERCISES FOR SECTION 4

(See *Answers to Practice Exercises for Section 4* on page **237**)

1) Use a spreadsheet to format a cell in different ways. Enter the number 0.34 in the first cell. Using the spreadsheet menu bar and online help, change the formatting of the cell to display this number as a percentage, a dollar amount, and then back to a general number.

2) Create a multiplication table. Enter the numbers 1 through 10 in the first column (A), and in the first row (1). In cell B2 enter the formula B1 × A2. In cell B3, enter the formula B1 × A3. Repeat this operation down the column. The last cell in column B should have the formula B1 × A10. For column C, formulas are C1 × A2, C1 × A3, C1 × A4, ..., C1 × A10. Complete entering formulas for columns C through J in the same manner. Use your spreadsheet to look up the value of 2×2, 5×7, and 8×9.

	A	B	C	D	E	F	G	H	I	J
1	1	2	3	4	5	6	7	8	9	10
2	2									
3	3									
4	4									
5	5									
6	6									
7	7									
8	8									
9	9									
10	10									

3) Write a single formula that can be entered into cell B2 of the previous problem and then copied into each blank cell (B2...J10) to complete the multiplication table (*Hint:* Use relative and absolute addressing). Explain how the formula changes as it is copied into different locations.

4) Use a spreadsheet to recreate **Table 1b** on page **29**. Make sure to format currency cells where required.

5) Using your spreadsheet's online help for guidance, recreate the compound interest calculation in **Table 2** on page **32** using the spreadsheet's future value interest rate function. Make sure to format currency and percentage cells correctly.

6) Using the spreadsheet you created in the previous question, calculate the future value of $2,500 compounded annually for 12 years at 7.5% interest. What would the future value be if the interest was compounded monthly?

7) An equation for windchill temperature is given on *Handbook* page **2884**. Build a windchill temperature table. What is the windchill temperature for: a) 15° F and 20 mph wind, b) 55° F and 20 mph wind, c) 5° F and 5 mph wind?

8) The windchill equation on *Handbook* page **2884** contains two sets of parentheses. Are these parentheses required when entering the formula into a spreadsheet? Why?

SECTION 5

CALCULATIONS INVOLVING LOGARITHMS

Machinery's Handbook page **31**

The purpose of logarithms is to facilitate and shorten calculations involving multiplication and division, obtaining the powers of numbers, and extracting the roots of numbers. By means of logarithms, long multiplication problems become simple addition of logarithms; cumbersome division problems are easily solved by simple subtraction of logarithms; the fourth root or, say, the 10.4th root of a number can be extracted easily; and any number can be raised to the twelfth power as readily as it can be squared.

The availability of inexpensive computers and handheld calculators has eliminated much of the need to use logarithms for such purposes; there are, however, many applications in which the logarithm of a number is used to obtain the solution of a problem. For example, in the section *Compound Interest* on page **139** of the *Handbook*, there is a formula to find the number of years n required for a sum of money to grow a specified amount. The example accompanying the formula shows the calculations that include the logarithms 3, 2.69897, and 0.025306, which correspond to the numbers 1000, 500, and 1.06, respectively. These logarithms were obtained directly from a handheld calculator and are the common or *Briggs* system of logarithms, which have a base 10. Any other system of logarithms, such as that of base e ($e = 2.71828\ldots$), could have been used with the same result. Base e logarithms are sometimes called *natural logarithms*.

There are other types of problems in which logarithms of a specific base, usually 10 or e, must be used to obtain the correct result. On the logarithm keys of most calculators, base 10 logs are identified by the word "log" and those of base e are referred to as "ln".

In the common or Briggs system of logarithms, which is used ordinarily, the base of the logarithms is 10; that is, the logarithm is the *exponent* that would be affixed to 10 to produce the number corresponding to the logarithm. To illustrate:

Logarithm of 10 = 1 because $10^1 = 10$

Logarithm of 100 = 2 because $10^2 = 100$

Logarithm of 1000 = 3 because $10^3 = 1000$

In each case, it will be seen that the exponent of 10 equals the logarithm of the number. The logarithms of all numbers between 10 and 100 equal 1 plus some fraction. For example: The logarithm of 20 = 1.301030.

The logarithms of all numbers between 100 and 1000 = 2 plus some fraction; between 1000 and 10,000 = 3 plus some fraction; and so on. The tables of logarithms in engineering handbooks give only this fractional part of a logarithm, which is called the *mantissa*. The whole number part of a logarithm, which is called the *characteristic*, is not given in the tables because it can easily be determined by simple rules. The logarithm of 350 is 2.544068. The whole number 2 is the characteristic (see *Handbook* page **32**) and the decimal part 0.544068, the mantissa, is found in the table (*Machinery's Handbook 32 Digital Edition* page **3175**). Logarithms can most easily be found by using a scientific calculator.

Principles Governing the Application of Logarithms.—When logarithms are used, the product of two numbers can be obtained as follows: Add the logarithms of the two numbers; the sum equals the logarithm of the product. For example: The logarithm of 10 (commonly abbreviated log 10) equals 1; log 100 = 2; 2 + 1 = 3, which is the logarithm of 1000 and the product of 100 × 10.

Logarithms would not be used for such a simple example of multiplication; these particular numbers are employed merely to illustrate the principle involved.

For division by logarithms, subtract the logarithm of the divisor from the logarithm of the dividend to obtain the logarithm of the quotient. To use another simple example, divide 1000 by 100 using logarithms. The respective logarithms of these numbers are 3 and 2, and the difference equals the logarithm of the quotient 10.

In using logarithms to raise a number to any power, simply multiply the logarithm of the number by the exponent of the number; the product equals the logarithm of the power. To illustrate, find the value of 10^3 using logarithms. The logarithm of 10 = 1 and the exponent is 3; hence, $3 \times 1 = 3$ = log of 1000; hence, $10^3 = 1000$.

To extract any root of a number, merely divide the logarithm of this number by the index of the root; the quotient is the logarithm of the root. Thus, to obtain the cube root of 1000, divide 3 (log 1000) by 3 (index of root); the quotient equals 1, which is the logarithm of 10. Therefore,

$$\sqrt[3]{1000} = 10$$

Logarithms are of great value in many engineering and shop calculations because they make it possible to readily solve cumbersome or difficult problems that otherwise would require complicated formulas or higher mathematics. Keep constantly in mind that logarithms are merely exponents. Any number might be the base of a system of logarithms. Thus, if 2 were selected as a base, then the logarithm of 256 would equal 8 because $2^8 = 256$. However, unless otherwise mentioned, the term "logarithm" is used to apply to the common or Briggs system, which has 10 for a base.

The tables of common logarithms are found on *Machinery's Handbook 32 Digital Edition* pages **3175** and **3176**. The natural logarithms, *Machinery's Handbook 32 Digital Edition* pages **3177** and **3178**, are based upon the number 2.71828. These logarithms are used in higher mathematics and also in connection with the formula to determine the mean effective pressure of steam in engine cylinders.

Finding the Logarithms of Numbers.—There is nothing complicated about the use of logarithms, but a little practice is required to readily locate the logarithm of a given number or to reverse this process and find the number corresponding to a given logarithm. These corresponding numbers are sometimes called *antilogarithms*.

Carefully study the rules for finding logarithms given on *Handbook* page **31** and in the section *Logarithms* starting on page **3169**, *Machinery's Handbook 32 Digital Edition*. Although the characteristic or whole-number part of a logarithm is easily determined, the following table will assist the beginner in memorizing the rules.

Sample Numbers and Their Characteristics

Number	Characteristic	Number	Characteristic
0.008	−3	88	1
0.08	−2	888	2
0.8	−1	8888	3
8.0	0	88888	4

For common logarithms: For numbers greater than or equal to 1, the characteristic is 1 less than the number of places to the left of the decimal point. For numbers smaller than 1 and greater than 0, the characteristic is negative and its numerical value is 1 more than the number of zeros immediately to the right of the decimal point.

Example 1: Use of the table of numbers and their characteristics.

What number corresponds to the log $\overline{2}.55145$? (Note that the bar over a digit is used when the characteristic is negative and the mantissa is positive.) Find 0.551450 in the log tables to correspond to 356. From the table of characteristics, note that a −2 characteristic calls for one zero in front of the first integer; hence, 0.0356 is the number corresponding to the log $\overline{2}.55145$. Evaluating logarithms with negative characteristics is explained more thoroughly later.

Example 2: Find the logarithm of 46.8.

The mantissa of this number is 0.670246. When there are two whole-number places, the characteristic is 1; hence, the log of 46.8 is 1.670246.

After a little practice with the above table, you will become familiar enough with the rules governing the characteristic so that referring to the table will no longer be necessary.

Obtaining More Accurate Values than Given Directly by Tables.—The method of using the tables of logarithms to obtain more accurate values than are given directly, by means of interpolation, is explained on *Machinery's Handbook 32 Digital Edition* page **3170**. These instructions should be read carefully in order to understand the procedure in connection with the following example:

Example 3:

$$\frac{76824 \times 52.076}{435.21} =$$

log 76824 = 4.88549	log numerator = 6.60213
log 52.076 = <u>1.71664</u>	− log 435.21 = <u>2.63870</u>
log numerator = 6.60213	log quotient = 3.96343

The number corresponding to the logarithm 3.96343 is 9192.4. The logarithms just given for the dividend and divisor are obtained by interpolation from the log table in the following manner:

In the log tables on *Machinery's Handbook 32 Digital Edition* page **3175**, find the mantissa corresponding to the first three digits of the number 76824 and the mantissa of the next higher 3-digit number in the table, 769. The mantissa of 76824 is the mantissa of 768 plus $^{24}/_{100}$ times the difference between the mantissas of 769 and 768.

$$\begin{aligned} \text{Mantissa } 769 &= .885926 \\ \text{Mantissa } 768 &= \underline{.885361} \\ \text{Difference} &= .000565 \end{aligned}$$

Thus, log 76824 = 0.24 × 0.000565 + log 76800 = 4.885497. The characteristic 4 is obtained as previously illustrated in the table on page **41**. By again using interpolation, as explained in the *Handbook*, the corrected mantissas are found for the logarithms of 52.076 and 435.21.

After obtaining the logarithm of the quotient, which is 3.96343, interpolation is again used to determine the corresponding number more accurately than would be possible otherwise. The mantissa .96343 (see *Machinery's Handbook 32 Digital Edition* page **3176**) is found in the table between 0.963316 and 0.963788, the mantissas corresponding to 919 and 920, respectively.

$$0.963788 - 0.963316 = 0.000472$$

$$0.96343 - 0.963316 = 0.000114$$

Note that the first line gives the difference between the two mantissas nearest .96343, and the second line gives the difference between the mantissa of the quotient and the nearest smaller mantissa in the *Handbook* table. The characteristic 3 in the quotient 3.96343 indicates 4 digits before the decimal point in the answer, and thus the number sought is 9190 + $^{114}/_{472}$ (9200 − 9190) = 9192.4.

Changing Form of Logarithm Having Negative Characteristic.— The characteristic is frequently rearranged for easier manipulation. Note that 8 – 8 is the same as 0; hence, the log of 4.56 could be stated as 0.658965 or 8.658965 – 8. Similarly, the log of 0.075 = $\bar{2}.875061$ or 8.875061 – 10 or 7.875061 – 9. Any similar arrangement could be made, as determined by case in multiplication or division.

Example 4:

$$\sqrt[3]{0.47} = ?$$

$$\log 0.47 = \bar{1}.672098 \text{ or } 8.672098 - 9$$

$$\log \sqrt[3]{0.47} = (8.672098 - 9) \div 3 = 2.890699 \div 3 = \bar{1}.89070$$

In the first line above, 9 – 9 was added to log 0.47 because 3 (the index of the root) will divide evenly into 9; 11 – 12 or 5 – 6 could have been used as well. (Refer also to **Example 7** on *Machinery's Handbook 32 Digital Edition* page **3172**. The procedure differs from that just described, but the same result is obtained.)

To find the number corresponding to $\bar{1}.89070$, locate the nearest mantissa. Mantissa .890421 is found in the table and corresponds to 777. The $\bar{1}$ characteristic indicates that the decimal point immediately precedes the first integer; therefore, the number equivalent to the log 1.89070 is 0.777. If desired, additional accuracy can be obtained by interpolation, as explained previously. Thus, $\sqrt[3]{0.47} = 0.777$.

Cologarithms.—The cologarithm of a number is the logarithm of the reciprocal of that number. The resulting *cologs* have no properties different from those of ordinary logarithms, but they enable division to be done by addition because the addition of a colog is the same as the subtraction of a logarithm.

Example 5: $\dfrac{742 \times 6.31}{55 \times 0.92} = ?$

Note that this problem could be stated as 742 × 6.31 × 1/55 × 1/0.92. Then the logs of each number could be added because the process is one of multiplication only.

Log 1/55 can be obtained readily in two ways:

$$\log 1/55 = \log 1 - \log 55$$

$$\begin{array}{rrl} \log 1 = & 10.000000 & -10 \\ -\log 55 = & -1.740363 & \\ \hline & 8.259637 & -10 \quad = \bar{2}.259637 \end{array}$$

or

$$\log 1/55 = \log 0.0181818 \text{ (see reciprocals)}$$

$$\log 0.0181818 = \bar{2}.25964$$

This number $\bar{2}.259637$ is called the colog of 55; hence, to find the colog of any number, subtract the logarithm of that number from 10.000000 – 10; this is the same as dividing 1 by the number whose colog is sought.

To find the colog of 0.92. subtract log 0.92 (or $\bar{1}.963788$) from 10.00000 – 10; thus:

$$\begin{array}{rrl} & 10.000000 & -10 \\ \log 0.92 = & \bar{1}.963788 & \\ \hline \text{colog } 0.92 = & 9.963788 & -10 \quad = 0.036212 \end{array}$$

(In subtracting negative characteristics, change the sign of the lower one and add.)

Another method is to use log 0.92 = $\bar{1}.96379$ or 9.96379 – 10, and proceed as above:

$$\begin{array}{llrl} & & 10.000000 & -10 \\ \log 0.92 = & \bar{1}.963788 = & 9.963788 & -10 \\ \hline \text{colog } 0.92 & = & 0.036212 & \end{array}$$

Example 5 may then be solved by adding logs; thus:

$$\begin{array}{lcr} \log 742 & = & 2.870404 \\ \log 6.31 & = & 0.800029 \\ \text{colog } 55 & = & \bar{2}.259637 \\ \text{colog } 0.92 & = & 0.036212 \\ \hline \text{log quotient} & = & 1.966282 \end{array}$$

The number corresponding to the logarithm of the quotient = 92.53.

Example 6: The initial absolute pressure of the steam in a steam engine cylinder is 120 pounds per square inch (psi); the length of the stroke is 26 inches; the clearance, $1\frac{1}{2}$ inches; and the period of admission, measured from the beginning of the stroke, 8 inches. Find the mean effective pressure.

The mean effective pressure is found by the formula:

$$p = \frac{P(1 + \log_e R)}{R}$$

in which p = mean effective pressure in pounds per square inch

P = initial absolute pressure in pounds per square inch

R = ratio of expansion, which in turn is found from the formula:

$$R = \frac{L + C}{l + C}$$

in which L = length of stroke in inches

l = period of admission in inches

C = clearance in inches

The given values are $P = 120$, $L = 26$, $l = 8$, and $C = 1$. By inserting the last three values in the formula for R, we have:

$$R = \frac{26 + 1\frac{1}{2}}{8 + 1\frac{1}{2}} = \frac{27.5}{9.5} = 2.89$$

We now insert the value of P and the found value of R in the formula for p:

$$p = \frac{120(1 + \log_e 2.89)}{2.89}$$

The logarithm may be found from tables or a calculator. The natural logarithm for 2.89 is 1.061257 (see *Machinery's Handbook 32 Digital Edition* page **3177**). Inserting this value in the formula, we have:

$$p = \frac{120(1 + 1.061257)}{2.89} = \frac{120 \times 2.061257}{2.89} = 85.6 \text{ lb/in}^2$$

PRACTICE EXERCISES FOR SECTION 5

(See *Answers to Practice Exercises for Section 5* on page **238**)

1) What are the rules governing the characteristics?

2) Find the mantissas of: 762, 478, 26, 0.0098, 6743, and 24.82.

3) What are the characteristics of the numbers just given?

4) What numbers could correspond to the following mantissas: 0.085016, 0.88508, and 0.22763?

5) (a) If the characteristic of each of the mantissas just given is 1, what would the corresponding numbers be? (b) Using the following characteristics (2, 0, 3) for each mantissa, find the antilogarithms or corresponding numbers.

6) $\log 765.4 = ?$; $\log 87.2 = ?$; $\log 0.00874 = ?$

7) What are the antilogarithms of: 2.89894, 1.24279, 0.18013, and 2.68708?

8) Find by interpolation the logarithm of: 75186 and 42.037.

9) Find the numbers corresponding to the following logarithms: 1.82997 and 0.67712.

10) $(2.71)^5 = ?$ $\quad (4.23)^{2.5} = ?$

11) $\sqrt{97.65} = ?$ $\qquad \sqrt[5]{4687} = ?$ $\qquad \sqrt[2.3]{44.5} = ?$

12) $\dfrac{62876 \times 54.2 \times 0.0326}{1728 \times 231} = ?$

13) $(2/19)^7 = ?$

14) $(9.16)^{2.47} = ?$

15) $\sqrt[3]{\dfrac{(75)^2 \times (5.23)^{2/3}}{0.00036 \times \sqrt{51.7}}} =$

16) The area of a circular sector = $0.008727ar^2$ where a = angle in degrees and r = radius of the circle. Find the area of a circular sector the radius of which is 6.25 inches and the central angle is 42° 15′.

17) The diameter of a lineshaft carrying pulleys may be found from the formula: $d = \sqrt[3]{53.5\text{hp/rpm}}$. Find the diameter of shafting necessary to transmit 50 hp at 250 rpm.

18) The horsepower of a steam engine is found from the formula: $hp = PLAN/33000$, where

P = mean effective pressure in pounds per square inch
L = length of stroke in feet
A = area of piston in square inches
N = number of strokes per minute = revolutions per minute ×2

Find the horsepower of a steam engine if the pressure is 120 pounds, stroke is 18 inches, piston is 10 inches in diameter, and the number of revolutions per minute is 125.

19) Can the tables of logarithms be used for addition and subtraction?

20) Can logarithms be used to solve gear-ratio problems?

SECTION 6

DIMENSIONS, AREAS, AND VOLUMES OF FIGURES

Machinery's Handbook pages **37** to **88**

The formulas given for the solution of different problems relating to the areas of surfaces and volumes of various geometrical figures are derived from plane and solid geometry. For purposes of shop mathematics, all that is necessary is to select the appropriate figure and use the formula given. Keep in mind the applicable tables that are provided and use them in the solution of the formulas whenever such usage can be done to advantage.

Many rules may be developed directly from the table for polygons on *Handbook* page **77**. These rules will permit easy solution of nearly every problem involving a regular polygon. For instance, in the first "*A*" columns at the left, $A/S^2 = 7.6942$ for a decagon (10-sided polygon); by transposition, $S = \sqrt{A \div 7.6942}$. In the first "*R*" column, $R = 1.3066S$ for an octagon (8-sided polygon); hence, $S = R \div 1.3066$.

Given the frequent occurrence of such geometrical figures as squares, hexagons, spheres, and spherical segments in shop calculations, such tables dealing with these figures are very useful.

Example 1: A rectangle 12 inches long has an area of 120 square inches; what is the length of its diagonal?

The area of a rectangle equals the product of the two sides; hence, the unknown side of this rectangle equals $^{120}/_{12} = 10$ inches.

$$\text{Length of diagonal} = \sqrt{12^2 + 10^2} = \sqrt{244} = 15.6205$$

Example 2: If the diameter of a sphere, the diameter of the base, and the height of a cone are all equal, find the volume of the sphere if the volume of the cone is 250 cubic inches.

The formula on *Handbook* page **85** for the volume of a cone shows that the value for $250 = 0.2618d^2h$, in which d = diameter of cone base and h = vertical height of cone; hence,

$$d^2 = \frac{250}{0.2618h}$$

Since in this example d and h are equal,

$$d^3 = \frac{250}{0.2618}$$

and

$$d = \sqrt[3]{\frac{250}{0.2618}} = 9.8474 \text{ inches}$$

By referring to the formula on *Handbook* page **86**, the volume of a sphere $= 0.5236d^3 = 0.5236 \times (9.8474)^3 = 500$ cubic inches.

In solving the following exercises, first, construct the figure carefully, and then apply the formula. Use the examples in the *Handbook* as models.

PRACTICE EXERCISES FOR SECTION 6

(See *Answers to Practice Exercises for Section 6* on page **239**)

1) Find the volume of a cylinder having a base radius of 12.5 and a height of 16.3 inches.

2) Find the area of a triangle with sides that are 12, 14, and 18 inches in length.

3) Find the volume of a torus or circular ring made from $1\frac{1}{2}$ inch round stock with an outside diameter of 14 inches.

4) A bar of hexagonal screw stock measures 0.750 inch per side. What is the largest diameter that can be turned from this bar?

5) Using the prismoidal formula (*Handbook* page **66**), find the volume of the frustum of a regular triangular pyramid if its lower base is 6 inches per side, upper base 2 inches per side, and height 3 inches. (Use the table on *Handbook* page **77** for areas. The side of the midsection equals one-half the sum of one side of the lower base and one side of the upper base.)

6) What is the diameter of a circle the area of which is equivalent to that of a spherical zone whose radius is 4 inches and height 2 inches?

7) Find the volume of a steel ball $^3/_8$ inch in diameter.

8) What is the length of the side of a cube if the volume equals the volume of a frustum of a pyramid with square bases, 4 inches and 6 inches per side, and 3 inches high?

9) Find the volume of a bronze bushing if its inside diameter is 1 inch, outside diameter is $1^1/_2$ inches, and length is 2 inches.

10) Find the volume of a hollow sphere with an outside diameter of 10 inches and an inside diameter of 6 inches.

11) Find the area of a 10-equal-sided polygon inscribed in a 6-inch diameter circle.

12) What is the radius of a fillet if its chord is 2 inches? What is its area?

13) Find the area of the conical surface and volume of a frustum of a cone if the diameter of its lower base is 3 feet, diameter of upper base 1 foot, and height 3 feet.

14) Find the total area of the sides and the volume of a triangular prism 10 feet high, with a base width of 8 feet.

15) The diagonal of a square is 16 inches. What is the length of its side?

16) How many gallons can be contained in a barrel having the following dimensions: height $2^1/_2$ feet, bottom diameter 18 inches, and bilge diameter 21 inches? (The sides are formed to the arc of a circle.)

17) Find the area of a sector of a circle if the radius is 8 inches and the central angle is 32 degrees.

18) Find the height of a cone if its volume is 17.29 cubic inches and the radius of its base is 4 inches.

19) Find the volume of a rectangular pyramid having a base 4×5 inches and height 6 inches.

20) Find the distance across the corners of both hexagons and squares when the distance across in each case is: $\frac{1}{2}$, $1\frac{5}{8}$, $3\frac{3}{10}$, 5, and 8.

21) The diagonal of one square is 2.0329 and of the other square is 4.6846. Find the lengths of the sides of both squares.

22) In measuring the distance over plugs in a die that has six $\frac{3}{4}$-inch holes equally spaced on a circle, what should be the micrometer reading over opposite plugs if the distance over alternate plugs is $4\frac{1}{2}$ inches?

23) To what diameter should a shaft be turned in order to mill on one end a hexagon 2 inches on a side? An octagon 2 inches on a side?

SECTION 7

GEOMETRICAL PROPOSITIONS AND CONSTRUCTIONS

Machinery's Handbook pages **56** to **65**

Geometry is the branch of mathematics that deals with the relations of lines, angles, surfaces, and solids. Plane geometry treats the relations of lines, angles, and surfaces in one plane only, and since this branch of geometry is of special importance in mechanical work, the various propositions or fundamental principles are given in the *Handbook*, as well as various problems or constructions. This information is particularly useful in mechanical drafting and in solving problems about measurement.

Example 1: A segment-shaped casting (see **Fig. 1**) has a chordal length of 12 inches, and the height of the chord is 2 inches; determine by the application of a geometrical principle the radius R of the segment.

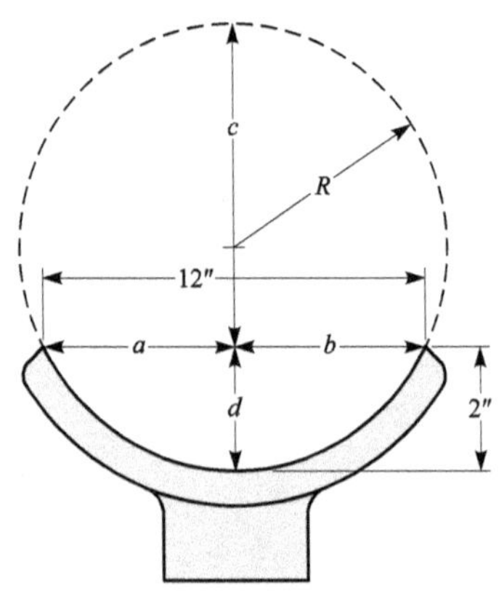

Fig. 1.

This problem may be solved by the application of the third geometrical proposition given on *Handbook* page **60**. In this example, one chord consists of two sections a and b, each 6 inches long; the other intersecting chord consists of one section d, 2 inches long; and the length of section c is to be determined in order to find radius R. Since $a \times b = c \times d$, it follows that:

$$c = \frac{a \times b}{d} = \frac{6 \times 6}{2} = 18 \quad \text{inches}$$

therefore,

$$R = \frac{c + d}{2} = \frac{18 + 2}{2} = 10 \quad \text{inches}$$

In this example, one chordal dimension, $c + d$ = the diameter; but the geometrical principle given in the *Handbook* applies regardless of the relative lengths of the intersecting chords.

Example 2: The center lines of three holes in a jig plate form a triangle. The angle between two of these intersecting center lines is 52 degrees. Another angle between adjacent center lines is 63 degrees. What is the third angle?

This problem is solved by application of the first geometrical principle on *Handbook* page **56**. The unknown angle = 180 – (63 + 52) = 65 degrees.

Example 3: The center lines of four holes in a jig plate form a four-sided figure. Three of the angles between the different intersecting center lines are 63 degrees, 105 degrees, and 58 degrees, respectively. What is the fourth angle?

According to the geometrical principle at the middle of *Handbook* page **58**, the unknown angle = 360 – (63 + 105 + 58) = 134 degrees.

Example 4: The centers of three holes are located on a circle. The angle between the radial center lines of the first and second holes is 22 degrees, and the center-to-center distance measured along the circle is $2\frac{1}{2}$ inches. The angle between the second and third holes is 44 degrees. What is the center-to-center distance along the circle?

This problem is solved by application of the fourth principle on *Handbook* page **60**. Since the lengths of the arcs are proportional to the angles, the center distance between the second and third holes = $(44 \times 2\frac{1}{2})/22 = 5$ inches. (See also rules governing proportion starting on *Handbook* page **9**.)

The following practice exercises relate to the propositions and constructions given and should be answered without the aid of the *Handbook*.

PRACTICE EXERCISES FOR SECTION 7

(See *Answers to Practice Exercises for Section 7* on page **239**)

1) If any two angles of a triangle are known, how can the third angle be determined?

2) State three instances where one triangle is equal to another.

3) When are triangles similar?

4) What is the purpose of proving triangles similar?

5) If a triangle is equilateral, what follows?

6) What are the properties of the bisector of any angle of an equilateral triangle?

7) What is an isosceles triangle?

8) How do the size of an angle and the length of a side of a triangle relate to each other?

9) Can you draw a triangle whose sides are 5, 6, and 11 inches?

10) What is the length of the hypotenuse of a right-angled triangle with sides measuring 12 and 16 inches?

11) What is the value of the exterior angle of any triangle?

12) What are the relations of angles formed by two intersecting lines?

13) Draw two intersecting straight lines and a circle tangent to these lines.

14) Construct a right-angled triangle given the hypotenuse and one side.

15) When are the areas of two parallelograms equal?

16) When are the areas of two triangles equal?

17) If a radius of a circle is perpendicular to a chord, what follows?

18) What is the relation between the radius and tangent of a circle?

19) What lines pass through the point of tangency of two tangent circles?

20) What are the attributes to two tangents drawn to a circle from an external point?

21) What is the value of an angle between a tangent and a chord drawn from the point of tangency?

22) Are all angles equal if their vertices are on the circumference of a circle and they are subtended by the same chord?

23) If two chords intersect within a circle, what is the value of the product of their respective segments?

24) How can a right angle be drawn using a semicircle?

25) What does the length of circular arcs in the same circle depend on?

26) What are the circumferences and areas of two circles proportional to?

SECTION 8

TRIGONOMETRY: FUNCTIONS OF ANGLES

Machinery's Handbook page **89** to **106**

The basis of trigonometry is proportion. If the sides of any angle are indefinitely extended and perpendiculars from various points on one side are drawn to intersect the other side, right triangles will be formed, and the ratios of the respective sides and hypotenuses will be identical. If the base of the smallest triangle thus formed is 1 inch, and the height is $\frac{1}{2}$ inch (see **Fig. 1**), the ratio between these sides is $1 \div \frac{1}{2} = 2$ or $\frac{1}{2} \div 1 = \frac{1}{2}$ depending upon how the ratio is stated. If the next triangle is measured, the ratio between the base and altitude will likewise be either 2 or $\frac{1}{2}$, and this will always be true for any number of triangles, if the angle remains unchanged. For example, $3 \div 1\frac{1}{2} = 2$ and $4\frac{1}{2} \div 2\frac{1}{4} = 2$ or $1\frac{1}{2} \div 3 = \frac{1}{2}$ and $2\frac{1}{4} \div 4\frac{1}{2} = \frac{1}{2}$.

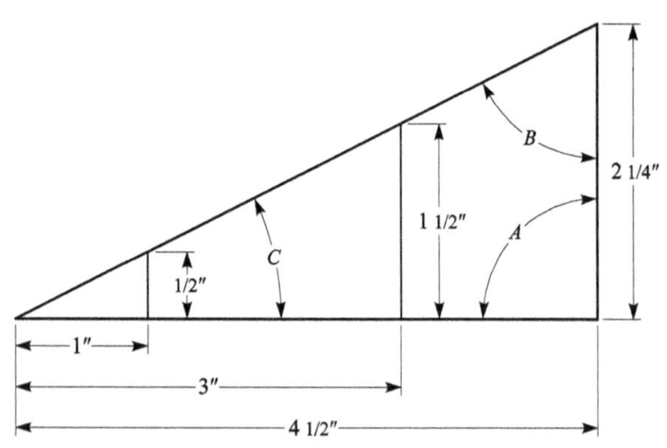

Fig. 1. For a Given Angle, the Ratio of the Base to the Altitude Is the Same for All Triangle Sizes

This relationship explains why rules can be developed to find the length of any side of a right triangle when the angle and one side are known or to find the angle when any two sides are known. Since there are two relations between any two sides of a right triangle, there can be, therefore, a total of six ratios with three sides. These ratios are defined and explained in the *Handbook*. Refer to pages **90** and **91** and note explanations of the terms *side adjacent*, *side opposite*, and *hypotenuse*.

The abbreviations of the trigonometric functions begin with a small letter and are not followed by periods.

Functions of Angles and Use of Trigonometric Tables.—On page **91** of the *Handbook* are the rules for determining the functions of angles. These rules, which should be memorized, may also be expressed as simple formulas:

$$\text{sine} = \frac{\text{side opposite}}{\text{hypotenuse}} \qquad \text{cosecant} = \frac{\text{hypotenuse}}{\text{side opposide}}$$

$$\text{cosine} = \frac{\text{side adjacent}}{\text{hypotenuse}} \qquad \text{secant} = \frac{\text{hypotenuse}}{\text{side adjacent}}$$

$$\text{tangent} = \frac{\text{side opposite}}{\text{side adjacent}} \qquad \text{cotangent} = \frac{\text{side adjacent}}{\text{side opposite}}$$

Note that these functions are arranged in pairs to include sine and cosecant, cosine and secant, tangent and cotangent, and that each pair consists of a function and its reciprocal. Also, note that these functions are merely ratios: the sine is the ratio of the *side opposite* to the *hypotenuse*, cosine is the ratio of the *side adjacent* to the *hypotenuse*, etc. Tables of trigonometric functions are, therefore, tables of ratios, and these functions can be obtained easily and quickly from most pocket calculators. For example, tan 20° 30′ = 0.37388; this means that in any right triangle having an acute angle of 20° 30′, the side opposite that angle is equal in length to 0.37388 times the length of the side adjacent. Cos 50° 22′ = 0.63787; this means that in any right triangle having an angle of 50° 22′, if the hypotenuse equals a certain length, say, 8, the side adjacent to the angle will equal 0.63787×8 or 5.10296.

Referring to **Fig. 1**, tan angle $C = 2\frac{1}{4} \div 4\frac{1}{2} = 1\frac{1}{2} \div 3 = \frac{1}{2} \div 1 = 0.5$; therefore, for this particular angle C, the *side opposite* is always equal to 0.5 times *side adjacent*, thus: $1 \times 0.5 = \frac{1}{2}$; $3 \times 0.5 = 1\frac{1}{2}$; and $4\frac{1}{2} \times 0.5 = 2\frac{1}{4}$. The side opposite angle B equals $4\frac{1}{2}$; hence, tan $B = 4\frac{1}{2} \div 2\frac{1}{4} = 2$.

Finding Angle Equivalent to Given Function.—After determining the tangent of angle C or of angle B, the values of these angles can be determined readily. As tan $C = 0.5$, find the numbers nearest to this in the tangent column on *Handbook* page **104**: 0.498582, corresponding to 26 degrees, 30 minutes, and 0.502219 corresponding to the angle 26 degrees, 40 minutes. Because 0.5 is approximately midway between 0.498582 and 0.502219, angle C can be accurately estimated as 26 degrees, 35 minutes. This degree of accuracy is usually sufficient; however, improved accuracy may be obtained by interpolation, as explained in the examples to follow.

Since angle $A = 90$ degrees, and, as the sum of three angles of a triangle always equals 180 degrees, it is evident that angle $C + B = 90$ degrees; therefore, $B = 90$ degrees minus 26 degrees, 35 minutes = 63 degrees, 25 minutes. The table on *Handbook* page **104** also shows that tan 63 degrees, 25 minutes is midway between 1.991164 and 2.005690, or approximately 2 within 0.0002.

Note that for angles 45° to 90°, *Handbook* pages **103–105**, the table is used by reading from the bottom up, using the function labels across the bottom of the table, as explained on *Handbook* page **101**.

In the foregoing example, the tangent is used to determine the unknown angles because the known sides are the side adjacent and the side opposite the unknown angles, which are the sides required for determining the tangent. If the side adjacent and the length of hypotenuse had been given instead, the unknown angles might have been determined by first finding the cosine because the cosine equals the side adjacent divided by the hypotenuse.

The acute angles (like B and C, **Fig. 1**) of any right triangle must be complementary, so the function of any angle equals the cofunction of its complement; thus, the sine of angle B = the cosine of angle C; the tangent of angle B = the cotangent of angle C; etc. Thus, $\tan B = 4\frac{1}{2} \div 2\frac{1}{4}$ and cotangent C also equals $4\frac{1}{2} \div 2\frac{1}{4}$. The tangent of 20° 30′ = 0.37388, which also equals the cotangent of 20° 30′. For this reason, it is only necessary to calculate the trigonometric ratios to 45° when making a table of trigonometric functions for angles between 45° and 90°, and this is why the functions of angles between 45° and 90° are located in the table by reading it backwards or in reverse order, as previously mentioned.

Example 1: Find the tangent of 44 degrees, 59 minutes.

Following instructions given on page **101** of the *Handbook*, find 44 degrees, 50 minutes, and 45 degrees, 0 minutes at the bottom of page **105**; and find their respective tangents, 0.994199 and 1.0000000, in the column labeled "tan" across the top of the table. The tangent of 44° 59′ is 0.994199 + 0.9 × (1 − 0.994199) = 0.99942.

Example 2: Find the tangent of 45 degrees, 5 minutes.

At the bottom of *Handbook* page **105**, and above "tan" at the bottom right of the table, are the tangents of 45° 0′ and 45° 10′, 1.000000 and 1.005835, respectively. The required tangent is midway between these two values and can be found from 1.000000 + 0.5 × (1.005835 − 1) = 1.00292.

How to Find More Accurate Functions and Angles than Those Given in the Table.—In the *Handbook*, the values of trigonometric functions are given for degrees and 10-minute increments; hence, if the given angle is in degrees, minutes, and seconds, the value of the function is determined from the nearest given values by interpolation.

Example 3: Determine the sine of 14° 22′ 26″. It is evident that this value lies between the sine of 14° 20′ and the sine of 14° 30′.

Sine 14° 20′ = 0.247563 and sine 14° 30′ = 0.250380; the difference = 0.250389 − 0.247563 = 0.002817. Consider this difference as a whole number (2817) and multiply it by a fraction with its numerator being the number of additional minutes and fractions of minutes (number of seconds divided by 60) in the given angle ($2 + {}^{26}\!/_{60}$), and its denominator being the number of minutes in the interval between 14° 20′ and the sine of 14° 30′.

Thus, $(2 + {}^{26}\!/_{60})/10 \times 2817 = [(2 \times 60) + 26]/(10 \times 60) \times 2817 =$ 685.47; hence, by adding 0.000685 to sine of 14° 20′, we find that sine 14° 22′ 26″ = 0.247563 + 0.000685 = 0.24825.

The correction value (represented in this example by 0.000685) is *added* to the function of the *smaller* angle nearest the given angle in dealing with sines or tangents, but this correction value is *subtracted* in dealing with cosines or cotangents.

Example 4: Find the angle whose cosine is 0.27052.

The table of trigonometric functions shows that the desired angle is between 74° 10′ and 74° 20′ because the cosines of these angles are, respectively, 0.272840 and 0.270040. The difference = 0.272840 – 0.270040 = 0.00280′. From the cosine of the smaller angle (i.e., the larger cosine) or 0.272840, subtract the given cosine; thus, 0.272840 – 0.27052 = 0.00232; hence 232/280 × 10 = 8.28571′ or the number of minutes to add to the smaller angle to obtain the required angle. Thus, the angle for a cosine of 0.27052 is 74° 18.28571′, or 74° 18′ 17″. Angles corresponding to given sines, tangents, or cotangents may be determined by the same method.

Trigonometric Functions of Angles Greater than 90 Degrees.—In obtuse triangles, one angle is greater than 90 degrees, and the *Handbook* tables can be used for finding the functions of angles larger than 90 degrees, but the angle must be first expressed in terms of an angle less than 90 degrees.

The sine of an angle greater than 90 degrees but less than 180 degrees equals the sine of an angle that is the difference between 180 degrees and the given angle.

Example 5: Find the sine of 118 degrees.

sin 118° = sin (180° – 118°) = sin 62°. Refer to page **104** to see that the sine given for 62 degrees is 0.882948.

The cosine, tangent, and cotangent of an angle greater than 90 but less than 180 degrees equals, respectively, the cosine, tangent, and cotangent of the difference between 180 degrees and the given angle; but the angular function has a negative value and must be preceded by a minus sign.

Example 6: Find tan 123 degrees, 20 minutes.

tan 123° 20′ = –tan (180° – 123° 20′) = –tan 56° 40′ = –1.520426

Example 7: Find csc 150 degrees.

Cosecant, abbreviated "csc" or "cosec," equals 1/sin, and is positive for angles 90 to 180 degrees (see *Handbook* page **102**):

$\csc 15° = 1/\sin(180° - 150°) = 1/\sin 30° = 1/0.5 = 2.0$

In the calculation of triangles, it is very important to include the minus sign in connection with the sines, cosines, tangents, and cotangents of angles greater than 90 degrees. *Signs of Trigonometric Functions, Fractions of π, and Degree-Radian Conversion* (**Fig. 4** on page **102** of the *Handbook*) shows clearly the negative and positive values of different functions and angles between 0 and 360 degrees. *Useful Relationships Among Angles* (**Table 1** on page **102**) is also helpful in determining the function, sign, and angle less than 90 degrees that is equivalent to the function of an angle greater than 90 degrees.

Use of Functions for Laying Out Angles.—Trigonometric functions may be used for laying out angles accurately either on drawings or in connection with template work, etc. The following example illustrates the general method:

Example 8: Construct or lay out an angle of 27 degrees, 29 minutes by using its sine instead of a protractor.

First, draw two lines at right angles, as in **Fig. 2**, and to any convenient length. Using a calculator, find the sine of 27 degrees, 29 minutes: 0.46149. If there is enough space, lay out your figure to an enlarged scale to obtain greater accuracy. Assume that the scale is 10 to 1: therefore, multiply the sine of the angle by 10, obtaining 4.6149 or about $4^{39}/_{64}$. Set the dividers or the compass to this dimension and with *a* (**Fig. 2**) as a center, draw an arc, thus obtaining one side of the triangle *ab*. Now set the compass to 10 inches (since the scale is 10 to 1) and, with *b* as the center, describe an arc so as to obtain intersection *c*. The hypotenuse of the triangle is now drawn through the intersections *c* and *b*, thus obtaining an angle *C* of 27 degrees, 29 minutes within fairly close limits. The angle *C*, laid out in this way, equals 27 degrees, 29 minutes because:

$$\frac{\text{side opposite}}{\text{hypotenuse}} = \frac{4.6149}{10} = 0.46149 = \sin 27°29'$$

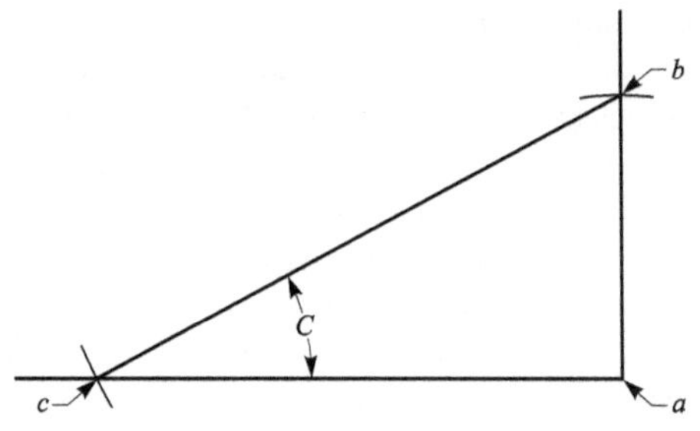

Fig. 2. Method of Laying Out Angle by Using Its Sine

Tables of Functions Used in Conjunction with Formulas.— When milling keyways, it is often desirable to know the total depth from the outside of the shaft to the bottom of the keyway. With this depth known, the cutter can be fed down to the required depth without taking any measurements other than that indicated by the graduations on the machine. To determine the total depth, it is necessary to calculate the height of the arc, which is designated as dimension *A* in **Fig. 3**. The formula usually employed to determine *A* for a given diameter of shaft *D* and width of key *W* is as follows:

$$A = \frac{D}{2} - \sqrt{\left(\frac{D}{2}\right)^2 - \left(\frac{W}{2}\right)^2}$$

Another formula, which is simpler than the one above, is used in conjunction with a calculator, as follows:

$$A = \frac{D}{2} \times \text{ versed sine of an angle whose sine is } \frac{W}{D}$$

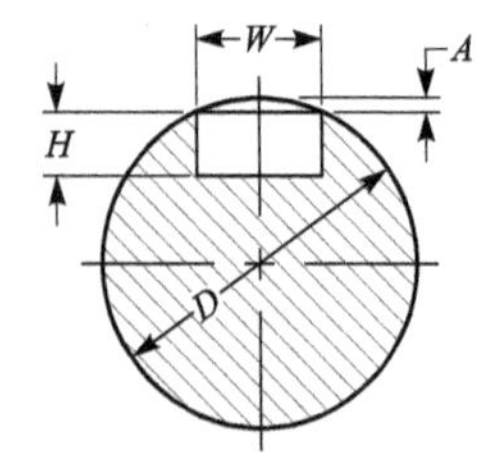

Fig. 3. To Find Height *A* for Arc of Given Radius and Width *W*

Example 9: To illustrate the application of this formula, let it be required to find the height A when the shaft diameter is $\frac{7}{8}$ inch and the width W of the key is $\frac{7}{32}$ inch. Then, $W/D = (\frac{7}{32})/(\frac{7}{8}) = \frac{7}{32} \times \frac{8}{7} = 0.25$. Using the formula at the bottom of *Handbook* page **106** for versed $\sin \theta = 1 - \cos \theta$, and a calculator, the angle corresponding to sin 0.25 = 14.4775 degrees, or 14 degrees, 28 minutes, 39 seconds. The cosine of this angle is 0.9682, and subtracting this value from 1 gives 0.03175 for the versed sine. Then, the height of the circular segment $A = D/2 \times 0.03175 = (7 \times 0.03175)/(8 \times 2) = 0.01389$, so the total depth of the keyway equals dimension H plus 0.01389 inch.

PRACTICE EXERCISES FOR SECTION 8

(See *Answers to Practice Exercises for Section 8* on page **240**)

1) How should a scientific pocket calculator be used to solve triangles?

2) Explain the meaning of $\sin 30° = 0.50000$.

3) Find $\sin 18° 26' 30''$, $\tan 27° 16' 15''$, and $\cos 32° 55' 17''$.

4) Find the angles that correspond to the following tangents: 0.52035 and 0.13025; to the following cosines: 0.06826 and 0.66330.

5) Give two rules for finding the *side opposite* a given angle.

6) Give two rules for finding the *side adjacent* to a given angle.

7) Explain the following terms: equilateral, isosceles, acute angle, obtuse angle, and oblique angle.

8) What is meant by complement, side adjacent, and side opposite?

9) Can the elements referred to in Exercise 8 be used in solving an isosceles triangle?

10) Construct by use of tangents an angle of 42° 20′.

11) Construct by use of sines an angle of 68° 15′.

12) Construct by use of cosines an angle of 55° 5′.

SECTION 9

SOLUTION OF RIGHT-ANGLE TRIANGLES

Machinery's Handbook pages **93–94**

A thorough knowledge of the solution of triangles or trigonometry is essential in drafting, layout work, bench work, and for convenient and rapid operation of some machine tools. Calculations concerning gears, screw threads, dovetails, angles, tapers, solution of polygons, gage design, cams, dies, and general inspection work are dependent upon trigonometry. Many geometrical problems may be solved more rapidly by trigonometry than by geometry.

In shop trigonometry, it is not necessary to develop and memorize the various rules and formulas, but it is essential that the six trigonometric functions be mastered thoroughly. Remember that a thorough, working knowledge of trigonometry depends upon drill work; hence a large number of problems should be solved.

The various formulas for the solution of right-angle triangles are given on *Handbook* page **93**; and examples showing their application are given on page **94**. These formulas may, of course, be applied to a large variety of practical problems in drafting rooms, tool rooms, and machine shops, as indicated by the following examples.

Whenever two sides of a right-angle triangle are given, the third side can always be found by a simple arithmetical calculation, as shown by the second and third examples on *Handbook* page **94**. To find the angles, however, it is necessary to use tables of sines, cosines, tangents, and cotangents, or a calculator, and, if only one side and one of the acute angles are given, the natural trigonometric functions (trig tables, pages **103–105**) must be used for finding the lengths of the other sides.

Example 1: The Jarno taper is 0.600 inch per foot for all numbers. What is the included angle?

As the angle measured from the axis or center line is 0.600 ÷ 2 = 0.300 inch per foot, the tangent of one-half the included angle = 0.300 ÷ 12 = 0.25 = tan 1° 26′; hence, the included angle = 2° 52′. A more direct method is to find the angle whose tangent equals the taper per foot divided by 24, as explained on *Handbook* page **699**.

Example 2: Determine the width W (see **Fig. 1**) of a cutter for milling a splined shaft having 6 splines 0.312 inch wide and a diameter B of 1.060 inches.

Dimension W may be computed by using the following formula:

$$W = \sin\left(\frac{\frac{360°}{N} - 2a}{2}\right) \times B$$

in which N = number of splines; B = diameter of body or of the shaft at the root of the spline groove.

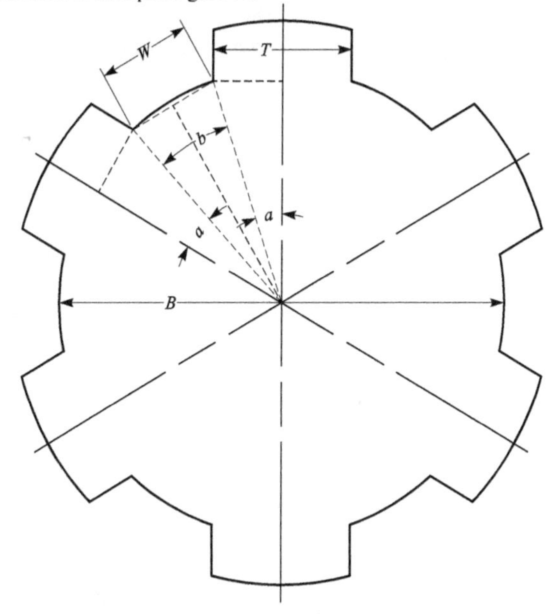

Fig. 1. To Find Width *W* of Spline-Groove Milling Cutter

Angle *a* must first be computed, as follows:

$$\sin a = \frac{T}{2} \div \frac{B}{2} \quad \text{or} \quad \sin a = \frac{T}{B}$$

where T = width of spline; B = diameter at the root of the spline groove. In this example,

$$\sin a = \frac{0.312}{1.060} = 0.29434$$

$$a = 17°7'; \text{ hence}$$

$$W = \left(\frac{\sin \frac{360°}{6} - 2 \times 17°7'}{2} \right) \times 1.060 = 0.236 \text{ inch}$$

This formula has also been used frequently in connection with broach design, but it is capable of a more general application. If the splines are to be ground on the sides, suitable deduction must be made from dimension W to leave sufficient stock for grinding.

If the angle b is known or is first determined, then

$$W = B \times \sin \frac{b}{2}$$

As there are 6 splines in this example, angle $b = 60° - 2a = 60° - 34° 14' = 25° 46'$; hence,

$W = 1.060 \times \sin 12° 53' = 1.060 \times 0.22297 = 0.236$ inch

Example 3: In sharpening the teeth of thread milling cutters, if the teeth have rake, it is necessary to position each tooth for the grinding operation so that the outside tip of the tooth is at a horizontal distance x from the vertical center line of the milling cutter as shown in **Fig. 2b**. What must this distance x be if the outside radius to the tooth tip is r and the rake angle is to be A? What distance x off center must a $4\frac{1}{2}$-inch diameter cutter be set if the teeth are to have a 3-degree rake angle?

In **Fig. 2**, it will be seen that, assuming the tooth has been properly sharpened to rake angle A, if a line is drawn extending the front edge of the tooth, it will be at a perpendicular distance x from the center of the cutter. Let the cutter now be rotated until the tip of the tooth is at a horizontal distance x from the vertical center line of the cutter as shown in **Fig. 2b**. It will be noted that an extension of the front edge of the cutter is still at perpendicular distance x from the center of the cutter, indicating that the cutter face is parallel to the vertical center line or is itself vertical, which is the desired position for sharpening using a vertical wheel. Thus, x is the proper offset distance for grinding the tooth to rake angle A if the radius to the tooth tip is r. Since r is the hypotenuse, and x is one side of a right-angled triangle,

$$x = r \sin A$$

For a cutter diameter of $4\frac{1}{2}$ inches and a rake angle of 3 degrees,

$$x = (4.5 \div 2) \sin 3° = 2.25 \times 0.05234$$
$$= 0.118 \text{ inch}$$

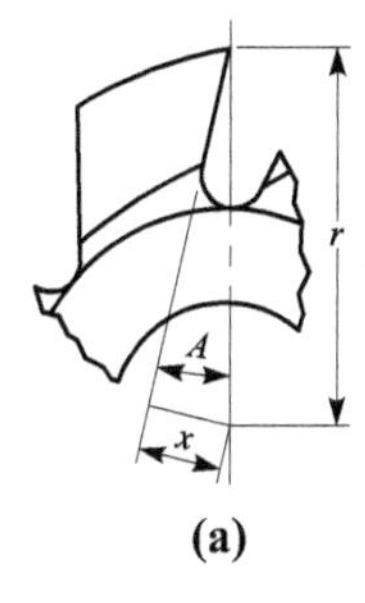

(a)

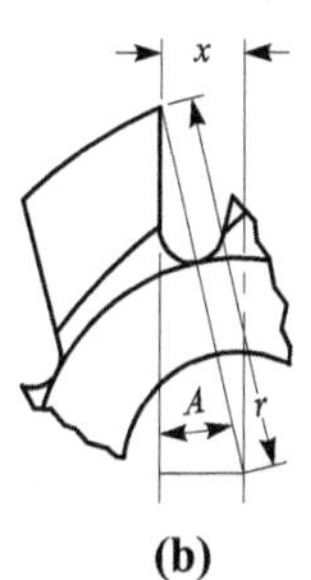

(b)

Fig. 2. To Find Horizontal Distance for Positioning Milling Cutter Tooth for Grinding Rake Angle *A*

Example 4: Forming tools are to be made for different sizes of poppet valve heads, and a general formula is required for finding angle x from dimensions given in **Fig. 3**.

The values for b, h, and r can be determined easily from the given dimensions. Angle x can then be found in the following manner:

Referring to the lower diagram in **Fig. 3**,

$$\tan A = \frac{h}{b} \quad (1)$$

$$c = \frac{h}{\sin A} \quad (2)$$

Also,

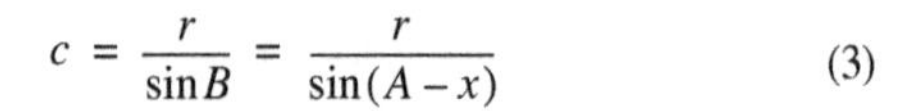

$$c = \frac{r}{\sin B} = \frac{r}{\sin(A - x)} \tag{3}$$

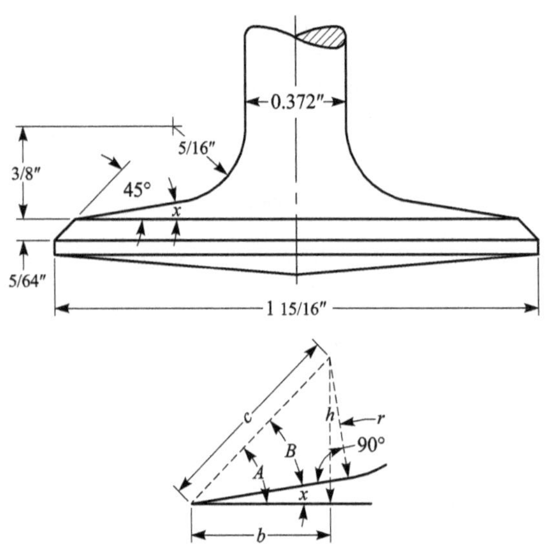

Fig. 3. To Find Angle *x*, Having the Dimensions Given on the Upper Diagram

From **Equations (2)** and **(3)** by comparison,

$$\frac{r}{\sin(A - x)} = \frac{h}{\sin A} \tag{4a}$$

$$\sin(A - x) = \frac{r \sin A}{h} \tag{4b}$$

From the dimensions given, it is obvious that $b = 0.392125$ inch, $h = 0.375$ inch, and $r = 0.3125$ inch. Substituting these values in **Equations (1)** and **(4b)** and solving, angle A will be found to be 43 degrees, 43 minutes, and angle $(A - x)$ to be 35 degrees, 10 minutes. By subtracting these two values, angle x will be found to equal 8 degrees, 33 minutes.

Example 5: In tool designing, it is frequently necessary to determine the length of a tangent to two circles. In **Fig. 4**, R = radius of large circle = $^{13}/_{16}$ inch; r = radius of small circle = $^{3}/_{8}$ inch; W = center distance between circles = $1^{11}/_{16}$ inches.

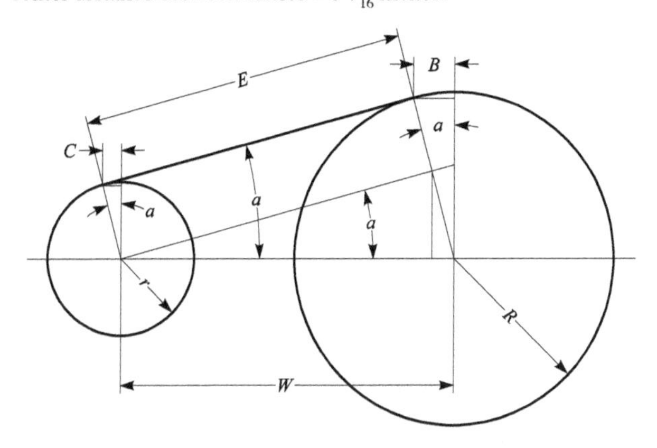

Fig. 4. To Find Dimension *E* or Distance Between Points of Tangency

With the values given, find the following: E = length of tangent, B = length of horizontal line from point of tangency on large circle to the vertical line, and C = length of horizontal line from point of tangency on small circle to the vertical center line.

$$\sin a = \frac{R-r}{W} = \frac{13/16 - 3/8}{1\frac{11}{16}} = 0.25925$$

$$\text{Angle } a = 15°1' \text{ nearly}$$

$$E = W\cos a = 1\frac{11}{16} \times 0.9658 = 1.63 \text{ inches}$$

$$B = R\sin a \qquad \text{and} \qquad C = r\sin a$$

Example 6: A circle is inscribed in a right triangle having the dimensions shown in **Fig. 5**. Find the radius of the circle.

In **Fig. 5**, $BD = BE$ and $AD = AF$, because tangents drawn to a circle from the same point are equal. $EC = CF$, and EC = radius OF. Then, let R = radius of inscribed circle. $AC - R = AD$ and $BC - R = DB$.

Adding,

$$AC + BC - 2R = AD + DB$$

$$AD + DB = AB$$

hence,

$$AC + BC - AB = 2R$$

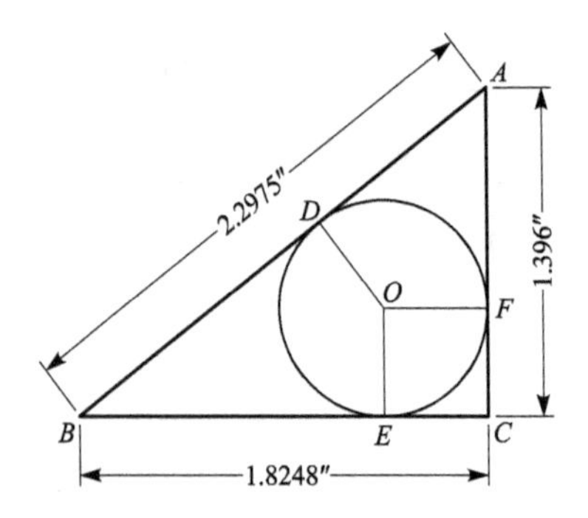

Fig. 5. To Find Radius of Circle Inscribed in Triangle

Stated as a rule: *The diameter of a circle inscribed in a right triangle is equal to the difference between the lengths of the hypotenuse and the sum of the lengths of the other sides*. Substituting the given dimensions, we have $1.396 + 1.8248 - 2.2975 = 0.9233 = 2R$ and $R = 0.4616$.

Example 7: A part is to be machined to an angle b of 30 degrees (**Fig. 6**) by using a vertical forming tool with a clearance angle a of 10 degrees. Calculate the angle of the forming tool as measured in a plane Z-Z, which is perpendicular to the front or clearance surface of the tool.

Assume that B represents the angle in plane Z-Z.

$$\tan B = \frac{Y}{X} \text{ and } Y = y \times \cos a \qquad (1)$$

Also,

$$y = X \times \tan b \text{ and } X = \frac{y}{\tan b} \qquad (2)$$

Now, substituting the values of Y and X in **Equation (1)**:

$$\tan B = \frac{y \times \cos a}{\frac{y}{\tan b}}$$

Clearing this equation of fractions,

$$\tan B = \cos a \times \tan b$$

In this example, $\tan B = 0.98481 \times 0.57735 = 0.56858$; hence, $B = 29°37'$ nearly.

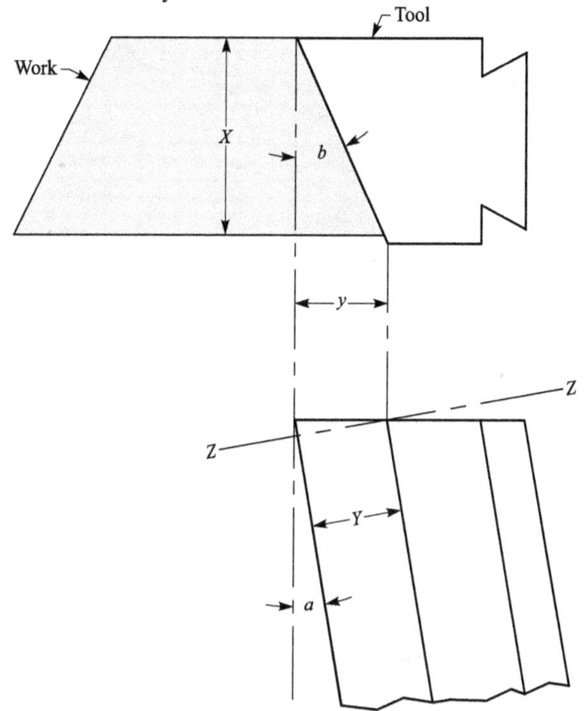

Fig. 6. The Problem is to Determine Angle of Forming Tool in Plane Z-Z

Example 8: A method of checking the diameter at the small end of a taper plug gage is shown by **Fig. 7**. The gage is first mounted on a sine-bar so that the top of the gage is parallel with the surface plate. A disk of known radius r is then placed in the corner formed by the end of the plug gage and the top side of the sine-bar. Now by determining the difference X in height between the top of the gage and the top edge of the disk, the accuracy of the diameter B can be checked readily. Derive formulas for determining dimension X.

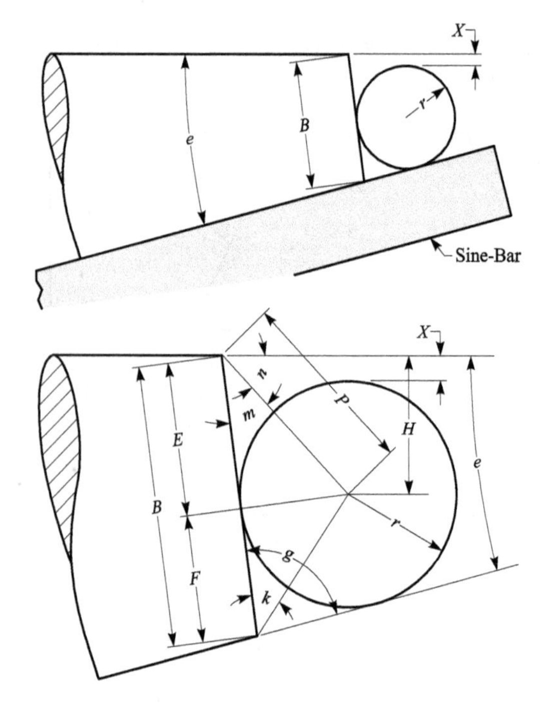

Fig. 7. The Problem is to Determine Height X in Order to Check Diameter B of Taper Plug

The known dimensions are:

e = angle of taper
r = radius of disk
B = required diameter at end of plug gage
$g = 90 \text{ degrees} - \frac{1}{2}e$
$k = \frac{1}{2}g$

By trigonometry,

$$F = \frac{r}{\tan k}; E = B - F; \text{and } \tan m = \frac{r}{E}$$

Also,

$$P = \frac{r}{\sin m}; n = g - m; \text{and } H = P \sin n$$

Therefore, $X = H - r$ or $r - H$, depending on whether or not the top edge of the disk is above or below the top of the plug gage. In **Fig. 7**, the top of the disk is below the top surface of the plug gage, so it is evident that $X = H - r$.

To illustrate the application of these formulas, assume that $e =$ 6 degrees, $r = 1$ inch, and $B = 2.400$ inches. The dimension X is then found as follows:

$$g = 90 - \tfrac{6}{2} = 87°; \text{and } k = 43° 30'$$

By trigonometry,

$$F = \frac{1}{0.9896} = 1.0538''; E = 2.400 - 1.0538 = 1.3462 \text{ inches}$$

$$\tan m = \frac{1}{1.3462} = 0.74283 \text{ and } m = 36° 36' 22''$$

$$P = \frac{1}{0.59631} = 1.6769''; \ n = 87° - 36°36'22'' = 50°23'38''$$

and

$$H = 1.6769 \times 0.77044 = 1.2920 \text{ inches}$$

Therefore,

$$X = H - r = 1.2920 - 1 = 0.2920 \text{ inch}$$

The disk here is below the top surface of the plug gage; hence, the formula $X = H - r$ was applied.

Example 9: In **Fig. 8**, $a = 1\tfrac{1}{4}$ inches, $h = 4$ inches, and angle $A =$ 12 degrees. Find dimension x and angle B.

Draw an arc through points E, F, and G, as shown, with r as a radius. According to a well-known theorem of geometry, which is given on *Handbook* page **59**, if an angle at the circumference of a circle, between two chords, is subtended by the same arc as the angle at the center, between two radii, then the angle at the circumference is equal to one-half the angle at the center. This being true, angle C is twice the magnitude of angle A, and angle $D =$ angle $A =$ 12 degrees.

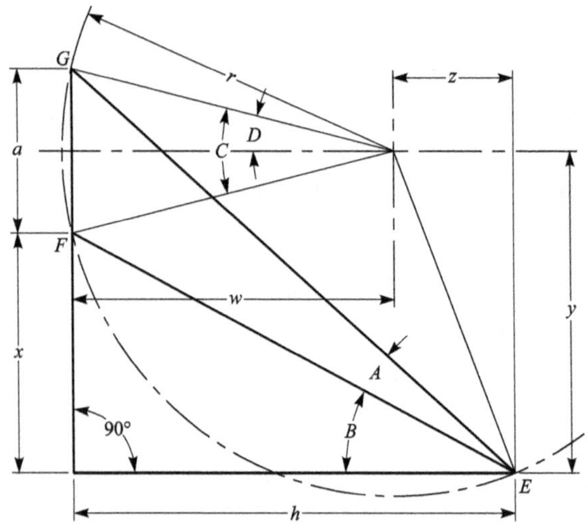

Fig. 8. Find Dimension *x* and Angle *B*, Given *a*, *h*, and Angle *A*

Thus,

$$r = \frac{a}{2\sin D} = \frac{1.25}{2 \times 0.20791} = 3.0061$$

$$w = \frac{a}{2}\cot D = 0.625 \times 4.7046 = 2.9404$$

and

$$z = h - w = 4 - 2.9404 = 1.0596$$

Now

$$y = \sqrt{r^2 - z^2} = \sqrt{7.9138505} = 2.8131$$

and

$$x = y - \frac{a}{2} = 2.8131 - 0.625 = 2.1881 \text{ inches}$$

Finally,

$$\tan B = \frac{x}{h} = \frac{2.1881}{4} = 0.54703$$

and

$$B = 28 \text{ degrees, } 40 \text{ minutes, } 47 \text{ seconds}$$

Example 10: A steel ball is placed inside a taper gage as shown in **Fig. 9**. If the angle of the taper, length of taper, radius of ball, and its position in the gage are known, how can the end diameters X and Y of the gage be determined by measuring dimension C?

The ball should be of such size as to project above the face of the gage. Although not necessary, this projection is preferable, as it permits the required measurements to be obtained more readily. After measuring the distance C, the calculation of dimension X is as follows: First obtain dimension A, which equals R multiplied by csc a. Then adding R to A and subtracting C, we obtain dimension B. Dimension X may then be obtained by multiplying $2B$ by the tangent of angle a. The formulas for X and Y can therefore be written as follows:

$$\begin{aligned} X &= 2(R\csc a + R - C)\tan a \\ &= 2(R\sec a + 2\tan a(R - C)) \\ Y &= X - 2T\tan a \end{aligned}$$

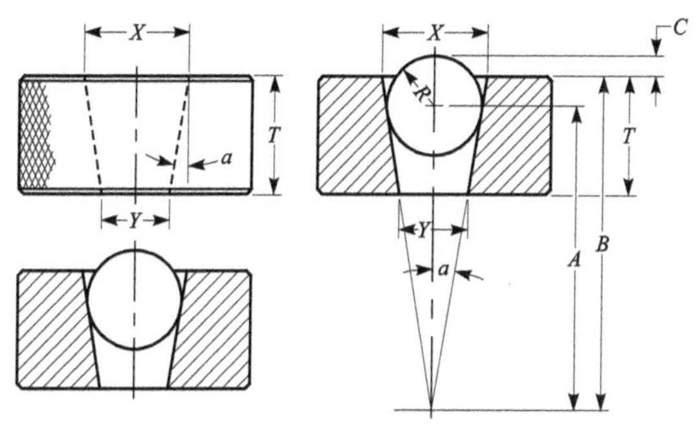

Fig. 9. Checking Dimensions X and Y by Using One Ball of Given Size

If, in **Fig. 9**, angle $a = 9$ degrees, $T = 1.250$ inches, $C = 0.250$ inch and $R = 0.500$ inch, what are the dimensions X and Y? Applying the formula,

$$X = 2 \times 0.500 \times 1.0125 + 2 \times 0.15838(0.500 - 0.250)$$

By solving this equation, $X = 1.0917$ inches. Then

$$Y = 1.0917 - (2.500 \times 0.15838) = 0.6957$$

Example 11: In designing a motion of the type shown in **Fig. 10**, it is essential, usually, to have link E swing equally above and below the center line M-M. A mathematical solution of this problem follows. In the illustration, G represents the machine frame; F, a lever shown in extreme positions; E, a link; and D, a slide. The distances A and B are fixed, and the problem is to obtain $A + X$, or the required length of the lever. In the right triangle:

$$A + X = \sqrt{(A - X)^2 + \left(\frac{B}{2}\right)^2}$$

Squaring, we have:

$$A^2 + 2AX + X^2 = A^2 - 2AX + X^2 + \frac{B^2}{4}$$

$$4AX = \frac{B^2}{4}$$

$$X = \frac{B^2}{16A}$$

$$A + X = A + \frac{B^2}{16A} = \text{length of lever}$$

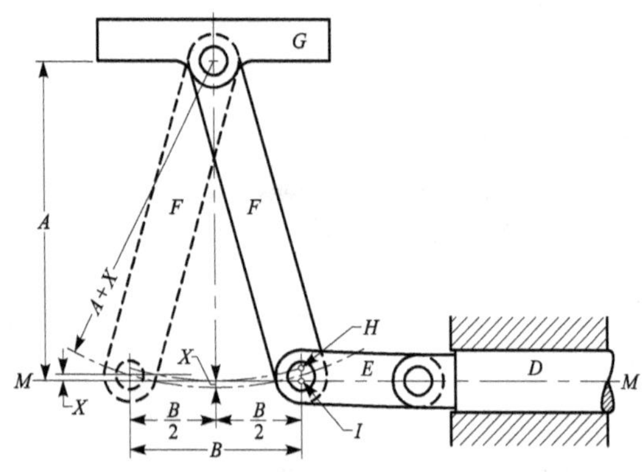

Fig. 10. Determining Length F so that Link E Will Swing Equally Above and Below the Center Line

To illustrate the application of this formula, assume that the length of a lever is required when the distance $A = 10$ inches, and the stroke B of the slide is 4 inches.

$$\text{Length of lever} = A + \frac{B^2}{16A} = 10 + \frac{16}{16 \times 10}$$
$$= 10.100 \text{ inches}$$

Thus, it is evident that the pin in the lower end of the lever will be 0.100 inch below the center line *M-M* when half the stroke has been made, and at each end of the stroke, the pin will be 0.100 inch above this center line.

Example 12: The spherical hubs of bevel gears are checked by measuring the distance x (**Fig. 11**) over a ball or plug placed against a plug gage that fits into the bore. Determine this distance x.

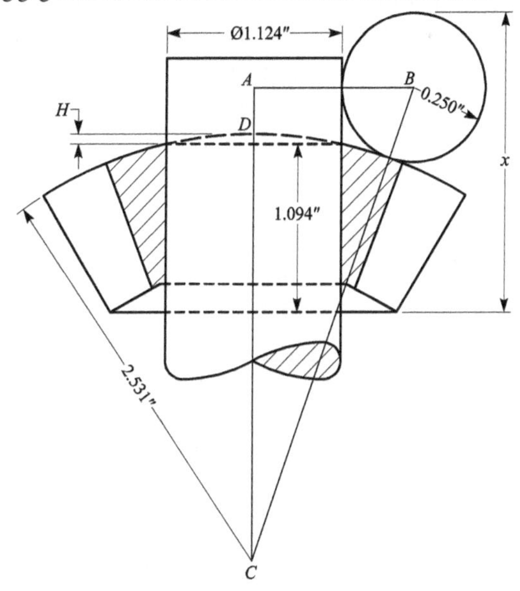

Fig. 11. Method of Checking the Spherical Hub of a Bevel Gear with Plug Gages

First find H by means of the formula for circular segments on *Handbook* page **74**.

$$H = 2.531 - 1/2\sqrt{4 \times (2.531)^2 - (1.124)^2} = 0.0632 \text{ inch}$$

$$AB = \frac{1.124}{2} + 0.25 = 0.812 \text{ inch}$$

$$BC = 2.531 + 0.25 = 2.781 \text{ inches}$$

Applying the formula on *Handbook* page **93** for a right triangle with side a and hypotenuse c known,

$$AC = \sqrt{(2.781)^2 - (0.812)^2} = 2.6599 \text{ inches}$$

$$AD = AC - DC = 2.6599 - 2.531 = 0.1289 \text{ inch}$$

$$x = 1.094 + 0.0632 + 0.1289 + 0.25 = 1.536 \text{ inches}$$

Example 13: The accuracy of a gage is to be checked by placing a ball or plug between the gage jaws and measuring to the top of the ball or plug as shown by **Fig. 12**. Dimension x is required, and the known dimensions and angles are shown by the illustration.

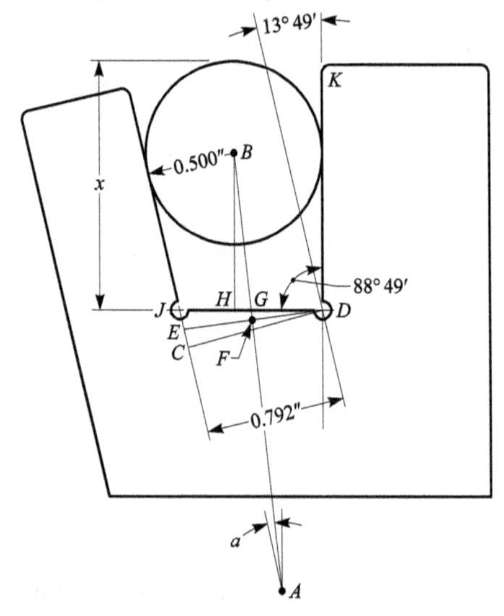

Fig. 12. Finding Dimension x to Check Accuracy of Gage

One-half of the included angle between the gage jaws equals one-half of $13° \times 49'$ or $6° \times 54\frac{1}{2}'$, and the latter equals angle a.

$$AB = \frac{0.500}{\sin 6°54\frac{1}{2}'} = 4.1569 \text{ inches}$$

DE is perpendicular to AB and angle CDE = angle a; hence,

$$DE = \frac{CD}{\cos 6°54\frac{1}{2}'} = \frac{0.792}{\cot 6°54\frac{1}{2}'} = 0.79779 \text{ inch}$$

$$AF = \frac{DE}{2} \times \cot 6°54\frac{1}{2}' = 3.2923 \text{ inches}$$

$$\text{Angle } CDK = 90° + 13°49' = 103°49'$$

$$\text{Angle } CDJ = 103°49' - 88°49' = 15°$$

$$\text{Angle } EDJ = 15° - 6°54\frac{1}{2}' = 8°5\frac{1}{2}'$$

$$GF = \frac{DE}{2} \times \tan 8°5\frac{1}{2}' = 0.056711 \text{ inch}$$

$$\text{Angle } HBG = \text{angle } EDJ = 8°5\frac{1}{2}'$$

$$BG = AB - (GF + AF) = 0.807889 \text{ inch}$$

$$BH = BG \times \cos 8°5\frac{1}{2}' = 0.79984 \text{ inch}$$

$$x = BH + 0.500 = 1.2998 \text{ inches}$$

If surface JD is parallel to the bottom surface of the gage, the distance between these surfaces might be added to x to make it possible to use a height gage from a surface plate.

Helix Angles of Screw Threads, Hobs, and Helical Gears.—The terms "helical" and "spiral" often are used interchangeably in drafting rooms and shops, although the two curves are entirely different. As the illustration on *Handbook* page **65** shows, every point on a helix is equidistant from the axis, and the curve advances at a uniform rate around a cylindrical area. The helix is illustrated by the springs shown on *Handbook* page **311**. A spiral is flat like a clock spring. A spiral may be defined mathematically as a curve having a constantly increasing radius of curvature.

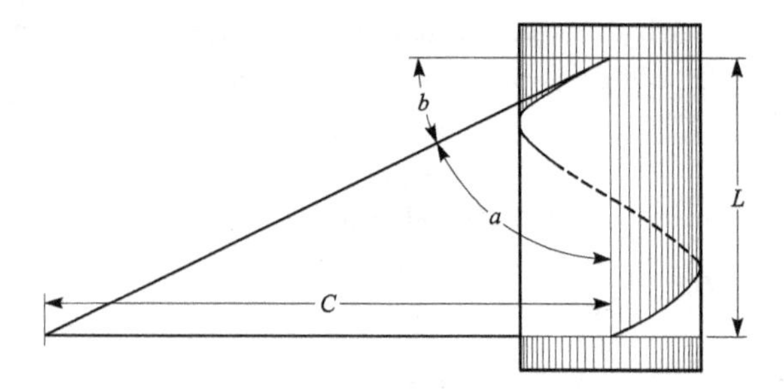

Fig. 13. Helix Represented by a Triangular Piece of Paper Wound Upon a Cylinder

If a piece of paper is cut in the form of a right triangle and wrapped around a cylinder (as indicated by **Fig. 13**), the hypotenuse will form a helix. The curvature of a screw thread represents a helix. From the properties of a right triangle, simple formulas can be derived for determining helix angles. Thus, if the circumference of a part is divided by the lead or distance that the helix advances axially in one turn, the quotient equals the tangent of the helix angle as measured from the axis. The angles of helical curves usually (but not always) are measured from the axis. The helix angle of a helical or spiral gear is measured from the axis, but the helix angle of a screw thread is measured from a plane perpendicular to the axis. In a helical gear, the angle is a (**Fig. 13**), whereas for a screw thread, the angle is b; hence, for helical gears, tan a of helix angle = C/L; for screw threads, tan b of helix angle = L/C. The helix angle of a hob, such as is used for gear cutting, also is measured as indicated at b and often is known as the *end angle*, because it is measured from the plane of the end surface of the hob. In calculating helix angles of helical gears, screw threads, and bobs, the pitch circumference is used.

Example 14: If the pitch diameter of a helical gear = 3.818 inches and the lead = 12 inches, what is the helix angle?

Tan helix angle = $(3.818 \times 3.1416) / 12 = 1$ very nearly; hence the angle = 45 degrees.

PRACTICE EXERCISES FOR SECTION 9

(See *Answers to Practice Exercises for Section 9* on page **241**)

1) The No. 4 Morse taper is 0.6233 inch per foot; calculate the included angle.

2) ANSI Standard pipe threads have a taper of $^3/_4$ inch per foot. What is the angle on each side of the center line?

3) To what dimension should the dividers be set to space 8 holes evenly on a circle of 6 inches diameter?

4) Explain the derivation of the formula

$$W = \sin\left(\frac{\frac{360^\circ}{N} - 2a}{2}\right) \times B$$

For notation, see **Example 2** and **Fig. 1** on page **65**.

5) The top of a male dovetail is 4 inches wide. If the angle is 55 degrees, and the depth is $^5/_8$ inch, what is the width at the bottom of the dovetail?

6) Angles may be laid out accurately by describing an arc with a radius of given length and then determining the length of a chord of this arc. In laying out an angle of 25 degrees, 20 minutes, using a radius of 8 inches, what should the length of the chord opposite the named angle be?

7) What is the largest square that may be milled on the end of a $2^1/_2$-inch bar of round stock?

8) A guy wire from a smoke stack is 120 feet long. How high is the stack if the wire is attached 10 feet from the top and makes an angle of 57 degrees with the stack?

9) In laying out a master jig plate, it is required that holes *F* and *H*, **Fig. 14**, shall be on a straight line that is $1^3/_4$ inch distant from hole *E*. The holes must also be on lines making, respectively, 40- and 50-degree angles with line *EG*, drawn at right angles to the sides of the jig plate through *E*, as shown in the figure. Find the dimensions $a, b, c,$ and d.

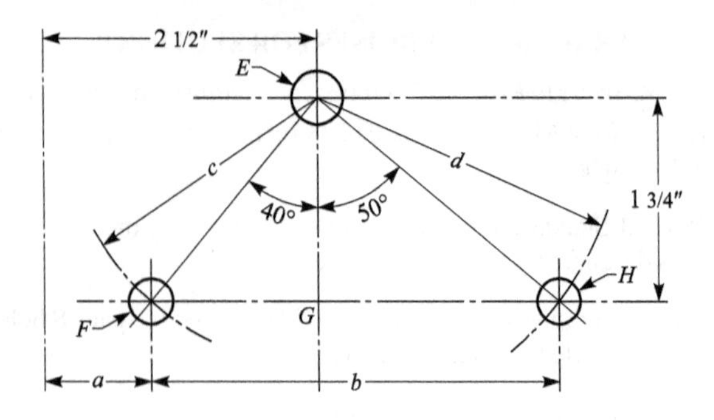

Fig. 14. Find Dimensions *a*, *b*, *c*, and *d*

10) **Fig. 15** shows a template for locating a pump body on a milling fixture, with the inside contour of the template corresponding with the contour of the pump flange. Find the angle *a* from the values given.

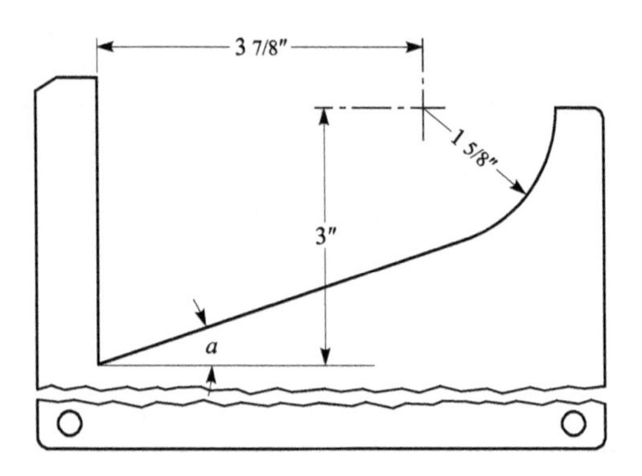

Fig. 15. Find an Angle *a* Having the Dimensions Given

11) Find the chordal distances as measured over plugs placed in holes located at different radii in the taximeter drive ring shown in **Fig. 16**. All holes are $\frac{7}{32}$ inch diameter; the angle between the center line of each pair of holes is 60 degrees.

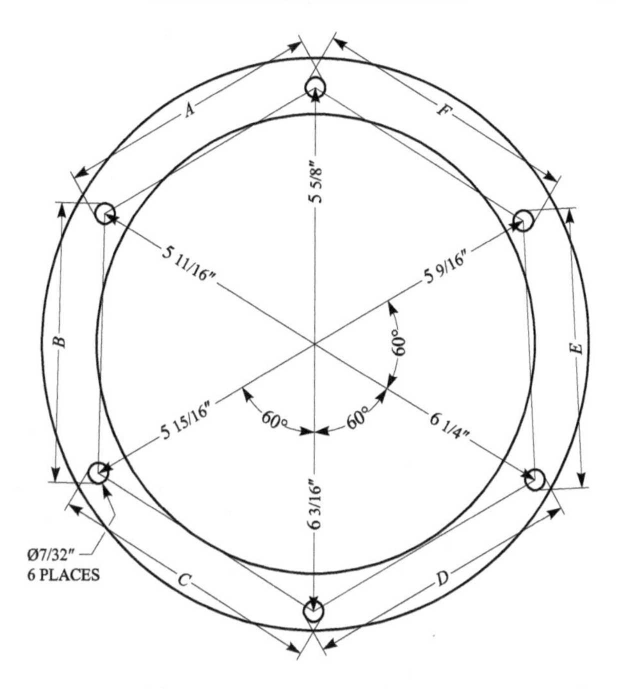

Fig. 16. To Find the Chordal Distances of Irregularly Spaced Holes Drilled in a Taximeter Drive Ring

12) An Acme screw thread has an outside diameter of $1\frac{1}{4}$ inches and has 6 threads per inch. Find the helix angle using the pitch diameter as a base. Also find the helix angle if a double thread is cut on the screw.

13) What is the lead of the flutes in a $\frac{7}{8}$-inch drill if the helix angle, measured from the center line of the drill, is 27° 30′?

14) A 4-inch diameter milling cutter has a lead of 68.57 inches. What is the helix angle measured from the axis?

SECTION 10

SOLUTION OF OBLIQUE TRIANGLES

Machinery's Handbook pages **95–97**

In solving problems for dimensions or angles, it is often convenient to work with oblique triangles. In an oblique triangle, none of the angles is a right angle. One of the angles may be over 90 degrees, or each of the three angles may be less than 90 degrees. Any oblique triangle may be solved by constructing perpendiculars to the sides from appropriate vertices, thus forming right triangles. The methods, previously explained, for solving right triangles will then solve the oblique triangles. The objection to this method of solving oblique triangles is that it is a long, tedious process.

Two of the examples in the *Handbook* on page **95**, which are solved by the formulas for oblique triangles, will be solved by the right-angle triangle method. These triangles have been solved to show that all oblique triangles can be solved thus and to give an opportunity to compare the two methods. There are four classes of oblique triangles:

1) Given one side and two angles

2) Given two sides and the included angle

3) Given two sides and the angle opposite one of them

4) Given the three sides

Example 1: Solve the first example on *Handbook* page **95** by the right-angle triangle method. By referring to **Fig. 1** on the next page:

$$\text{Angle } C = 180° - (62° + 80°) = 38°$$

Draw a line DC perpendicular to AB.

In the right triangle $BDC, DC/BC = \sin 62°$.

$$\frac{DC}{5} = 0.88295; DC = 5 \times 0.88295 = 4.41475$$

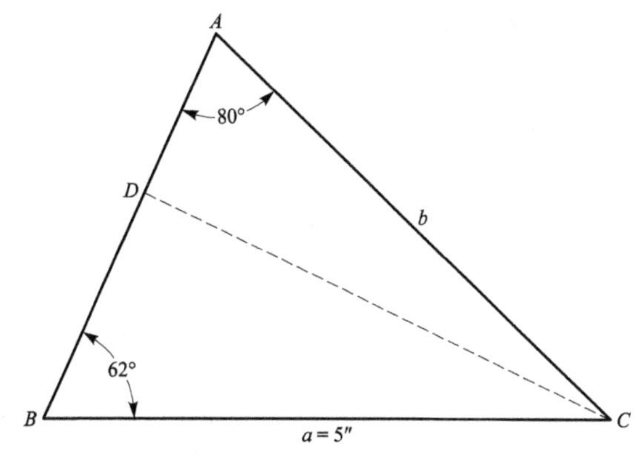

Fig. 1. Oblique Triangle Solved by Right-Angle Triangle Method

Angle $BCD = 90° - 62° = 28°; DCA = 38° - 28° = 10°$

$$\frac{BD}{5} = \cos 62°;\ BD = 5 \times 0.46947 = 2.34735$$

In triangle $ADC, AC/DC = \sec 10°$.

$$AC = 4.41475 \times 1.0154 = 4.4827$$

$$\frac{AD}{4.41475} = \tan 10°;\ AD = 4.41475 \times 0.17633 = 0.7785$$

$$\text{and } AB = AD + BD = 0.7785 + 2.34735 = 3.1258$$

$$C = 38°;\ b = 4.4827;\ c = 3.1258$$

Example 2: Apply the right-angle triangle method to the solution of the second example on *Handbook* page **95**.

Referring to **Fig. 2**, draw a line BD perpendicular to CA.

In the right triangle $BDC, BD/9 = \sin 35°$.

$$BD = 9 \times 0.57358 = 5.16222$$

$$\frac{CD}{9} = \cos 35°;\ CD = 9 \times 0.81915 = 7.37235$$

$$DA = 8 - 7.37235 = 0.62765$$

In the right triangle $BDA, \frac{BD}{DA} = \frac{5.16222}{0.62765} = \tan A$.

$\tan A = 8.2246$ and $A = 83°4'$

$$B = 180° - (83°4' + 35°) = 61°56'$$

$$\frac{BA}{BD} = \frac{BA}{5.1622} = \csc 83°4'; \qquad BA = 5.1622 \times 1.0074 = 5.2004$$

$$BA = 5.1622 \times 1.0074 = 5.2004$$

$$A = 83°4'; B = 61°56'; C = 35°$$

$$a = 9; b = 8; c = 5.2004$$

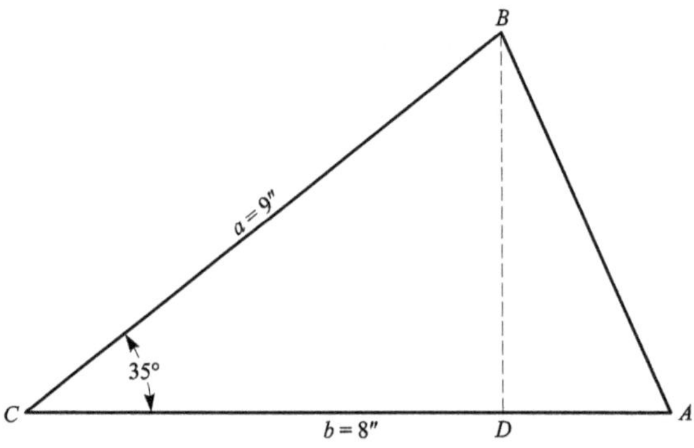

Fig. 2. Another Example of the Right-Angle Triangle Solution of an Oblique Triangle Equation

Use of Formulas for Oblique Triangles.—Oblique triangles are not encountered as frequently as right triangles, and, therefore, the methods of solving the latter may be fresh in the memory whereas methods for solving the former may be forgotten. All the formulas involved in the solution of the four classes of oblique triangles are derived from (1) the law of sines, (2) the law of cosines, and (3) the sum of angles of a triangle equal 180°.

The law of sines is that, in any triangle, the lengths of the sides are proportional to the sines of the opposite angles. (See diagrams on *Handbook* page **95**.)

$$\frac{a}{\sin A} = \frac{b}{\sin B} = \frac{c}{\sin C} \tag{1}$$

Solving this equation, we get:

$\frac{a}{\sin A} = \frac{b}{\sin B}$; then $a \sin B = b \sin A$ and

$a = \frac{b \sin A}{\sin B}$; $\sin B = \frac{b \sin A}{a}$

$b = \frac{a \sin B}{\sin A}$; $\sin A = \frac{a \sin B}{b}$

In like manner, $\frac{a}{\sin A} = \frac{c}{\sin C}$ and

$a \sin C = c \sin A$; hence $\sin A = \frac{a \sin C}{c}$

and $\frac{b}{\sin B} = \frac{c}{\sin C}$ or $b \sin C = c \sin B$

Thus, twelve formulas may be derived. As a general rule, only **Formula (1)** is remembered, and special formulas are derived from it as required.

The law of cosines states that, in any triangle, the square of any side equals the sum of the squares of the other two sides minus twice their product multiplied by the cosine of the angle between them. These relations are stated as the following formulas:

$$a^2 = b^2 + c^2 - 2bc \cos A \quad \text{or} \quad a = \sqrt{b^2 + c^2 - 2bc \cos A} \tag{1}$$

$$b^2 = a^2 + c^2 - 2ac \cos B \quad \text{or} \quad b = \sqrt{a^2 + c^2 - 2ac \cos B} \tag{2}$$

$$c^2 = a^2 + b^2 - 2ab \cos C \quad \text{or} \quad c = \sqrt{a^2 + b^2 - 2ab \cos C} \tag{3}$$

By solving **(1)**, $a^2 = b^2 + c^2 - 2bc \cos A$ for $\cos A$,

$$2bc \cos A = b^2 + c^2 - a^2 \quad \text{(transposing)}$$

$$\cos A = \frac{b^2 + c^2 - a^2}{2bc}$$

In like manner, formulas for $\cos B$ and $\cos C$ may be found.

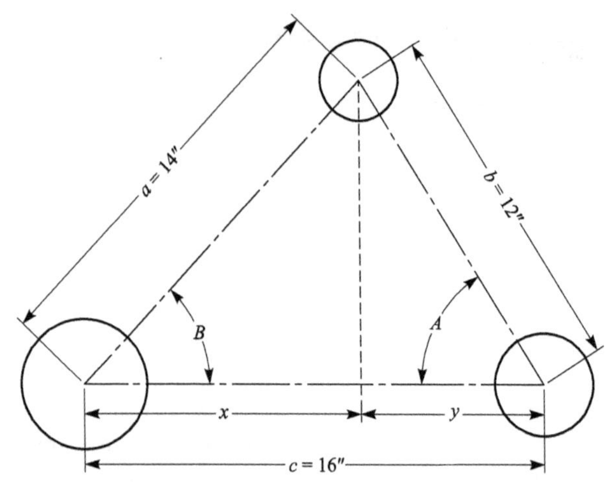

Fig. 3. Finding Dimensions *x* and *y* from Known Values

Example 3: A problem quite often encountered in layout work is illustrated in **Fig. 3**. Find the dimensions x and y between the holes, with these dimensions being measured from the intersection of the perpendicular line with the center line of the two lower holes. The three center-to-center distances are the only known values.

The first method might be to find the angle A (or B) by using a formula such as:

$$\cos A = \frac{b^2 + c^2 - a^2}{2bc}$$

and then solve the right triangle for y by the formula

$$y = b \cos A$$

Formulas (1) and **(2)** can be combined as follows:

$$y = \frac{b^2 + c^2 - a^2}{2c}$$

The value of x can be determined in a similar manner.

The second solution of this problem involves the following geometrical proposition: In any oblique triangle where the three sides are known, the ratio of the length of the base to the sum of the other two sides equals the ratio of the difference between the length of the two sides to the difference between the lengths x and y. Therefore, if $a = 14, b = 12,$ and $c = 16$ inches, then

$$c:(a+b) = (a-b):(x-y)$$
$$16:26 = 2:(x-y)$$
$$(x-y) = \frac{26 \times 2}{16} = 3\tfrac{1}{4} \text{ inches}$$

$$x = \frac{(x+y)+(x-y)}{2} = \frac{16+3\tfrac{1}{4}}{2} = 9.625 \text{ inches}$$

$$y = \frac{(x+y)-(x-y)}{2} = \frac{16-3\tfrac{1}{4}}{2} = 6.375 \text{ inches}$$

When Angles Have Negative Values.—In the solution of oblique triangles with one angle larger than 90 degrees, it is sometimes necessary to use angles whose functions are negative. (Review *Handbook* pages **3** and **129**.) Notice that for angles between 90 and 180 degrees, the cosine, tangent, cotangent, and secant are negative.

Example 4: In **Fig. 4**, two sides and the angle between them are shown. Find angles A and B. (See *Handbook* page **95**.)

$$\tan A = \frac{4 \times \sin 20°}{3 - 4 \times \cos 20°} = \frac{4 \times 0.34202}{3 - 4 \times 0.93969} = \frac{1.36808}{3 - 3.75876}$$

In the denominator of the fraction above, the number to be subtracted from 3 is greater than 3; the difference will thus be negative. Hence:

$$\tan A = \frac{1.36808}{3 - 3.75876} = \frac{1.36808}{-0.75876} = -1.80305$$

The final result is negative because a positive number (1.36808) is divided by a negative number (−0.75876). The tangents of angles greater than 90 degrees and smaller than 180 degrees are negative. To illustrate an angle whose tangent is negative, enter the value −1.80305 in the calculator and find the corresponding angle, which is −60.986558 degrees or −60 degrees, 59 minutes, 59 seconds. Because the tangent is negative, angle A must be subtracted from 180 degrees, giving 119.01344 degrees or 119 degrees, 0 minutes, 49 seconds as the angle. Now angle B is found from the formula,

$$B = 180° - (A + C) = 180° - (119°0'11'' + 20°)$$
$$= 180° - 139°0'11'' = 40°59'49''$$

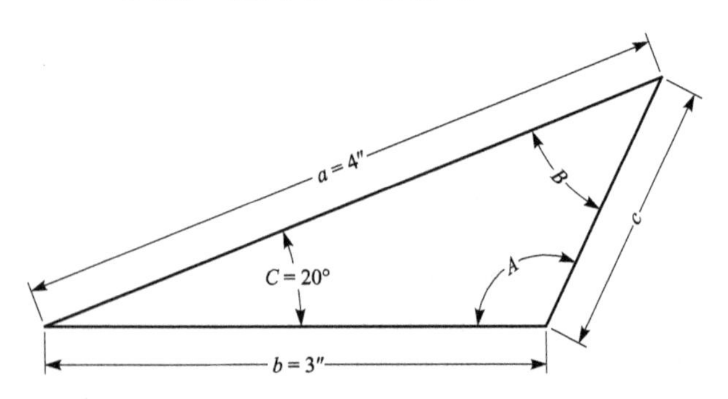

Fig. 4. Finding Angles *A* and *B* from the Dimensions Given

When Either of Two Triangles Conforms to the Given Dimensions.— When two sides and the angle opposite one of the given sides are known, *if the side opposite the given angle is shorter than the other given side*, two triangles can be drawn, having sides of the required length (as shown by **Fig. 5**) and the required angle opposite one of the sides. The lengths of the two known sides of each triangle are 8 and 9 inches, and the angle opposite the 8-inch side is 49 degrees, 27 minutes in each triangle; but it will be seen that the angle B of the lower triangle is very much larger than the corresponding angle of the upper triangle, and there is a great difference in the area. When two sides and one of the opposite angles are given, the problem can have two possible solutions when (and only when) the side opposite the given angle is shorter than the other given side. When the triangle to be calculated is drawn to scale, it is possible to determine from the shape of the triangle which of the two solutions applies.

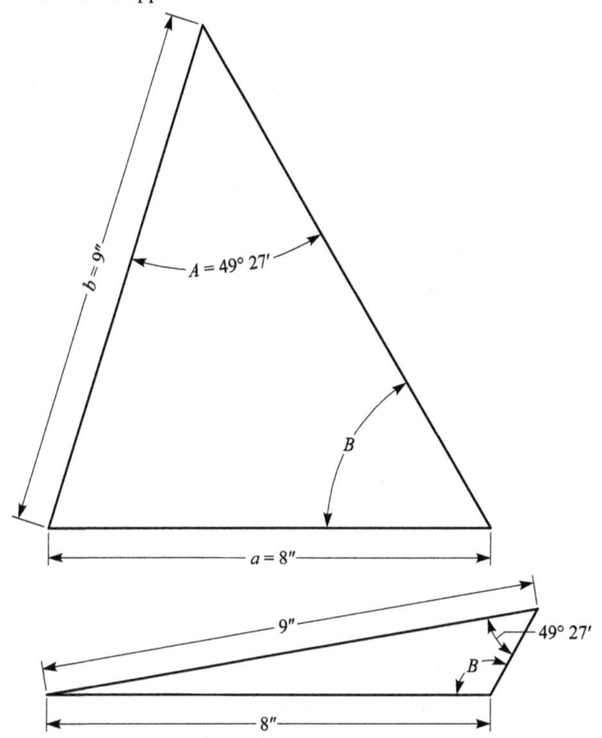

Fig. 5. Example of Two Possible Solutions of the Same Problem, Which Is to Find Angle *B*

Example 5: Find angle B, from the formula $\sin B = (b \sin A)/a$, where $b = 9$ inches; $A = 49$ degrees, 27 minutes; and a is the side opposite angle $A = 8$ inches.

$$\sin B = 9 \times 0.75984/8 = 0.85482 = \sin 58°44'34'' \quad \text{or}$$
$$\sin B = 121°15'36''$$

The practical requirements of the problem doubtless will indicate which of the two triangles shown in **Fig. 5** is the correct one.

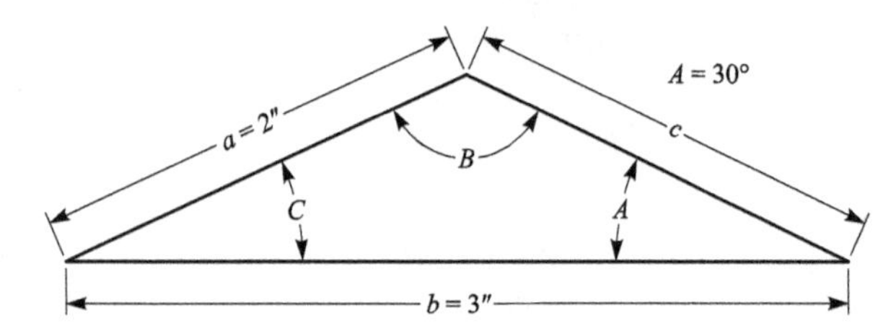

Fig. 6. Another Example that Has Two Possible Solutions

Example 6: In **Fig. 6**, $a = 2$ inches, $b = 3$ inches, and $A = 30$ degrees. Find B.

$$\sin B = \frac{b \times \sin A}{a} = \frac{\sin 30°}{2} = 0.75000$$

We find from the calculator that 0.75000 is the sine of 48°35′. From **Fig. 6** it is apparent, however, that B is greater than 90 degrees, and as 0.75000 is the sine not only of 48°35′, but also of 180° − 48°35′ = 131°25′, then angle B in this triangle equals 131°25′.

This example illustrates how the practical requirements of the problem indicate which of two angles is correct.

PRACTICE EXERCISES FOR SECTION 10

(See *Answers to Practice Exercises for Section 10* on page **242**)

1) Three holes in a jig are located as follows:

Hole No. 1 is 3.375 inches from hole No. 2 and 5.625 inches from hole No. 3; the distance between No. 2 and No. 3 is 6.250 inches. What three angles between the center lines are thus formed?

2) In **Fig. 7**, a triangle has one side measuring 6.5 feet, and two angles A and C are 78 and 73 degrees, respectively. Find angle B, sides b and c, and the area.

3) In **Fig. 8**, side a equals 3.2 inches; angle A, 118 degrees; and angle B, 40 degrees. Find angle C, sides b and c, and the area.

4) In **Fig. 9**, side b = 0.3 foot, angle B = 35°40′, and angle C = 24°10′. Find angle A, sides a and c, and the area.

5) Give two general rules for finding the areas of triangles.

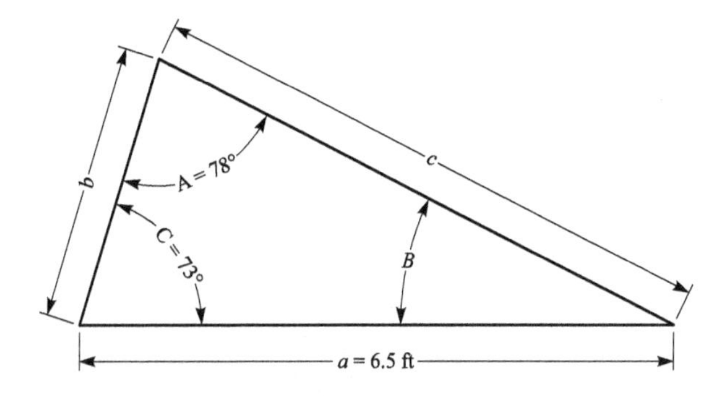

Fig. 7. Example for Practice Exercise No. 2

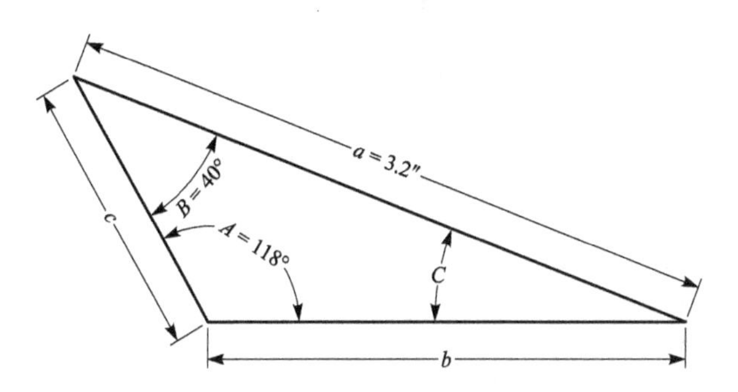

Fig. 8. Example for Practice Exercise No. 3

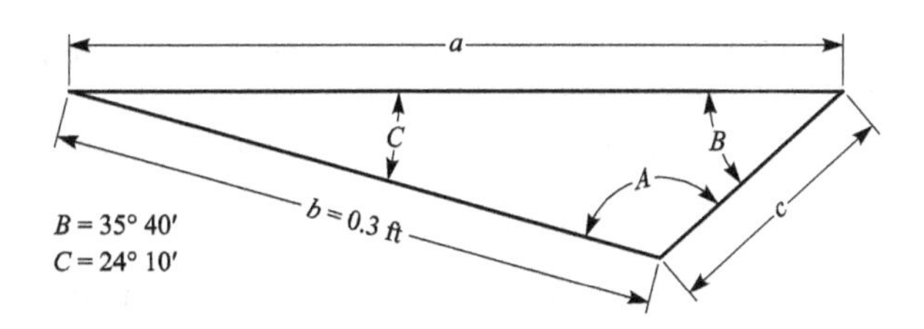

Fig. 9. Example for Practice Exercise No. 4

SECTION 11

FIGURING TAPERS

Machinery's Handbook pages **696–701**

The term "taper," as applied in shops and drafting rooms, means the difference between the large and small dimensions where the increase in size is uniform. Since tapering parts generally are conical, taper means the difference between the large and small diameters. Taper is ordinarily expressed as a certain number of inches per foot; thus, $\frac{1}{2}''$ per ft; $\frac{3}{4}''$ per ft; etc. In certain kinds of work, taper is also expressed as a decimal part of an inch per inch, as: 0.050″ per inch. The length of the work is always measured parallel to the center line (axis) of the work, and never along the tapered surface.

Suppose that the diameter at one end of a tapering part is 1 inch, and the diameter at the other end, 1.5 inches, and that the length of the part is 1 foot. This piece, then, tapers $\frac{1}{2}$ inch per foot because the difference between the diameters at the ends is $\frac{1}{2}$ inch. If the diameters at the ends of a part are $\frac{7}{16}$ inch and $\frac{1}{2}$ inch, and the length is 1 inch, this piece tapers $\frac{1}{16}$ inch per inch. The usual problems met when figuring tapers may be divided into seven classes. The rule to be used is found on *Handbook* page **699**.

Example 1: The diameter at the large end of a part is $2\frac{5}{8}$ inches, the diameter at the small end, $2\frac{3}{16}$ inches, and the length of the work, 7 inches. Find the taper per foot.

By referring to the third rule on *Handbook* page **699**,

$$\text{Taper per foot} = \frac{2\frac{5}{8} - 2\frac{3}{16}}{7} \times 12 = \frac{3}{4}\text{ inch}$$

Example 2: The diameter at the large end of a tapering part is $1\frac{5}{8}$ inches, the length is $3\frac{1}{2}$ inches, and the taper is $\frac{3}{4}$ inch per foot. The problem is to find the diameter at the small end.

By applying the fourth rule on *Handbook* page **699**,

$$\text{Diameter at small end} = 1\tfrac{5}{8} - \left(\frac{3/4}{12} \times 3\tfrac{1}{2}\right) = 1\tfrac{13}{32}$$

Example 3: What is the length of the taper if the two end diameters are 2.875 inches and 2.542 inches, the taper being 1 inch per foot?

By applying the sixth rule on *Handbook* page **699**,

$$\text{Distance between the two diameters} = \frac{2.875 - 2.542}{1} \times 12$$

$$= 4 \text{ inches nearly}$$

Example 4: If the length of the taper is 10 inches, and the taper is $\tfrac{3}{4}$ inch per foot, what is the taper in the given length?

By applying the last rule on *Handbook* page **699**,

$$\text{Taper in given length} = \frac{3/4}{12} \times 10 = 0.625 \text{ inch}$$

Example 5: The small diameter is 1.636 inches, the length of the work is 5 inches, and the taper is $\tfrac{1}{4}$ inch per foot; what is the large diameter?

By referring to the fifth rule on *Handbook* page **699**,

$$\text{Diameter at large end} = \left(\frac{1/4}{12} \times 5\right) + 1.636 = 1.740 \text{ inches}$$

Example 6: Sketch *A*, **Fig. 1**, shows a part used as a clamp bolt. The diameter, $3\tfrac{1}{4}$ inches, is given 3 inches from the large end of the taper. The total length of the taper is 10 inches. The taper is $\tfrac{3}{8}$ inch per foot. Find the diameter at the large and small ends of the taper.

First find the diameter of the large end using the fifth rule on *Handbook* page **699**.

$$\text{Diameter at large end} = \left(\frac{3/8}{12} \times 3\right) + 3\tfrac{1}{4} = 3\tfrac{11}{32} \text{ inches}$$

To find the diameter at the small end, use the fourth rule on *Handbook* page **699**.

$$\text{Diameter at small end} = 3\tfrac{11}{32} - \left(\frac{3/8}{12} \times 10\right) = 3\tfrac{1}{32} \text{ inches}$$

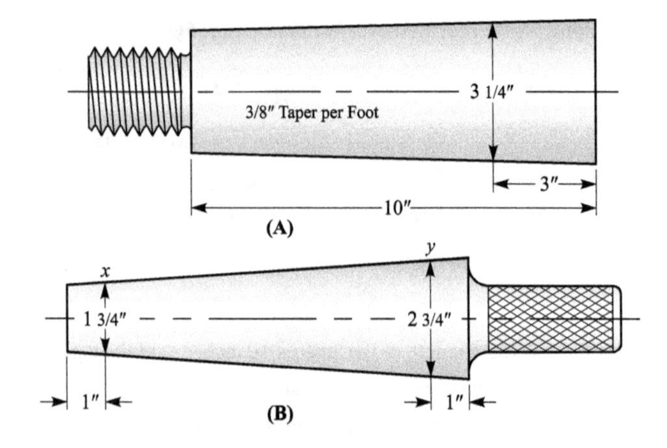

Fig. 1. Illustrations for Example 6 and Example 7

Example 7: At B, **Fig. 1**, is shown a taper master gage intended for inspecting taper-ring gages of various dimensions. The smallest diameter of the smallest ring gage is $1\frac{3}{4}$ inches, and the largest diameter of the largest ring gage is $2\frac{3}{4}$ inches. The taper is $1\frac{1}{2}$ inches per foot. It is required that the master gage extend 1 inch outside of the ring gages at both the small and the large ends, when these ring gages are tested. How long should the taper be on the master gage?

The sixth rule on *Handbook* page **699** may be applied here.

$$\text{Distance between the two diameters} = \frac{2\frac{3}{4} - 1\frac{3}{4}}{1\frac{1}{2}} \times 12$$

$$= 8 \text{ inches}$$

$$\text{Total length of taper} = 8 + 2 = 10 \text{ inches}$$

Table for Converting Taper per Foot to Degrees.—Some types of machines, such as milling machines, are graduated in degrees, making it necessary to convert the taper per foot to the corresponding angle in degrees. This conversion is quickly done by means of the table, *Handbook* page **698**.

Example 8: If a taper of $1\frac{1}{2}$ inches per foot is to be milled on a piece of work, at what angle must the machine table be set if the taper is measured from the axis of the work?

By referring to the table on *Handbook* page **698**, the angle corresponding to a taper of $1\frac{1}{2}$ inches to the foot is 3° 34′ 35″ as measured from the center line.

Note that the taper per foot varies directly as *the tangent of one-half the included angle.* Two mistakes frequently made in figuring tapers are assuming that the taper per foot varies directly as the included angle or that it varies directly as the tangent of the included angle. In order to verify this point, refer to the table on *Handbook* page **698**, where it will be seen that the included angle for a taper of 4 inches per foot (18° 55′ 29″) is not twice the included angle for a taper of 2 inches per foot (9° 31′ 38″). Neither is the tangent of 18° 55′ 29″ (0.3428587) twice the tangent of 9° 31′ 38″ (0.1678311).

Tapers for Machine-Tool Spindles.—The holes in machine-tool spindles, for receiving tool shanks, arbors, and centers, are tapered to ensure a tight grip, accuracy of location, and to facilitate removal of arbors, cutters, etc. The most common tapers are the Morse, the Brown & Sharpe, and the Jarno. The Morse has been very generally adopted for drilling-machine spindles. Most engine-lathe spindles also have the Morse taper, but some lathes have the Jarno or a modification of it, and others, a modified Morse taper, which is longer than the standard. A standard milling-machine spindle was adopted in 1927 by the milling-machine manufacturers of the National Machine Tool Builders' Association. A comparatively steep taper of $3\frac{1}{2}$ inches per foot was adopted in connection with this standard spindle to ensure instant release of arbors. Prior to the adoption of the standard spindle, the Brown & Sharpe taper was used for practically all milling machines and is also the taper for dividing-head spindles. There is considerable variation in grinding machine spindles. The Brown & Sharpe taper is the most common, but the Morse and the Jarno have also been used. Tapers of $\frac{5}{8}$ inch per foot and $\frac{3}{4}$ inch per foot also have been used to some extent on miscellaneous classes of machines requiring a taper hole in the spindle.

PRACTICE EXERCISES FOR SECTION 11

(See *Answers to Practice Exercises for Section 11* on page **242**)

1) What tapers, per foot, are used with the following tapers: a) Morse taper; b) Jarno taper; c) milling-machine spindle; d) and taper pin?

2) What is the taper per foot on a part if the included angle is 10° 30′; 55° 45′?

3) In setting up a taper gage like that shown on *Handbook* page **697**, what should be the center distance between 1.75-inch and 2-inch disks to check either the taper per foot or angle of a No. 4 Morse taper?

4) If it is required to check an angle of $14\frac{1}{2}$°, using two disks in contact, and the smaller disk is 1-inch diameter, what should the diameter of the larger disk be?

5) What should be the center distance, using disks of 2-inch and 3-inch diameter, to check an angle of 18° 30′ if the taper is measured from one side?

6) In grinding a reamer shank to fit a standard No. 2 Morse taper gage, it was found that the reamer stopped $\frac{3}{8}$ inch short of going into the gage to the gage mark. How much should be ground off the diameter?

7) A milling machine arbor has a shank $6\frac{1}{2}$ inches long with a No. 10 Brown & Sharpe taper. What is the total taper in this length?

8) A taper bushing for a grinding machine has a small inside diameter of $\frac{7}{8}$ inch. It is 3 inches long with $\frac{1}{2}$-inch taper per toot. Find the large inside diameter.

9) If a 5-inch sine-bar is used for finding the angle of the tapering bloc *A* (**Fig. 2**), and the heights of the sine-bar plug are as shown, find the corresponding angle *a* by means of the instructions beginning on *Handbook* page **694**.

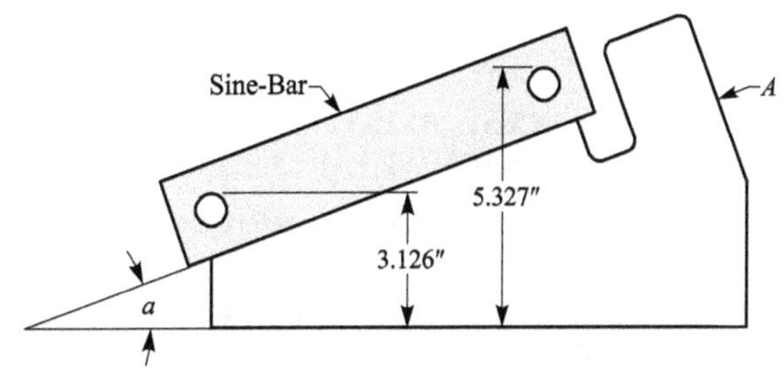

Fig. 2. Finding Angle *a* by Means of a Sine-Bar and Handbook Instructions

SECTION 12

TOLERANCES AND ALLOWANCES FOR MACHINE PARTS

Machinery's Handbook pages **641–690**

In manufacturing machine parts according to modern methods, certain maximum and minimum dimensions are established, particularly for the more important members of whatever machine or mechanism is to be constructed. These limiting dimensions serve two purposes: they prevent both unnecessary accuracy and excessive inaccuracies. A certain degree of accuracy is essential to the proper functioning of the assembled parts of a mechanism, but it is useless and wasteful to make parts more precise than needed to meet practical requirements. Hence, the use of proper limiting dimensions promotes efficiency in manufacturing and ensures standards of accuracy and quality that are consistent with the functions of the different parts of a mechanical device.

Parts made to specified limits usually are considered interchangeable or capable of use without selection, but there are several degrees of interchangeability in machinery manufacture. Strictly speaking, interchangeability consists of making the different parts of a mechanism so uniform in size and contour that each part of a certain model will fit any mating part of the same model, regardless of the lot to which it belongs or when it was made. However, as often defined, interchangeability consists in making each part fit any mating part in a certain series; that is, the interchangeability exists only in the same series. Selective assembly is sometimes termed interchangeability, but it involves a selection or sorting of parts as explained later. It will be noted that the strict definition of interchangeability does not imply that the parts must always be assembled without handwork, although that is usually considered desirable. It does mean, however, that when whatever process finishes the mating parts, they must assemble and function properly without fitting individual parts one to the other.

When a machine having interchangeable parts has been installed, possibly at some distant point, a broken part can readily be replaced by a new one sent by the manufacturer, but this feature is secondary as compared with the increased efficiency in manufacturing on an interchangeable basis. To make parts interchangeable, it is necessary to use gages and measuring tools, to provide some system of inspection, and to adopt suitable tolerances. Whether absolute interchangeability is practicable or not may depend upon the tolerances adopted, the relation between the different parts, and their form.

Meanings of "Limit," "Tolerance," and "Allowance"—The terms "limit" and "tolerance" and "allowance" are often used interchangeably, but each of these three terms has a distinct meaning and refers to different dimensions. As shown by **Fig. 1**, the *limits* of a hole or shaft are its diameters. *Tolerance* is the difference between two *limits* and limiting dimensions of a given part, and the term means that a certain amount of error is tolerated for practical reasons. *Allowance* is the difference between limiting dimensions on mating parts that are to be assembled either loosely or tightly, depending upon the amount allowed for the fit.

Example 1: Limits and fits for cylindrical parts are given starting on page **647** in the *Handbook*. These data provide a series of standard types and classes of fits. From **Table 8a** on page **654**, establish limits of size and clearance for a 2-inch diameter hole and shaft for a class RC-1 fit (hole H5, shaft g4).

$$\begin{aligned}
\text{Max. hole} &= 2 + 0.0005 = 2.0005\\
\text{Min. hole} &= 2 - 0 = 2\\
\text{Max. shaft} &= 2 - 0.0004 = 1.9996\\
\text{Min. shaft} &= 2 - 0.0007 = 1.9993\\
\text{Min. allow.} &= \text{min. hole} - \text{max. shaft} = 2 - 1.996 = 0.0004\\
\text{Max. allow.} &= \text{max. hole} - \text{min. shaft}\\
&= 2.0005 - 1.9993 = 0.0012
\end{aligned}$$

Example 2: Beginning on *Handbook* page **1952**, there are tables of dimensions for the Standard Unified Screw Thread Series—Class 1A, 2A, and 3A and B Fits. Determine the pitch-diameter tolerance of both screw and nut and the minimum and maximum allowance between screw and nut at the pitch diameter, assuming that the nominal diameter is 1 inch, the pitch is 8 threads per inch, and the fits are Class 2A and 2B for screw and nut, respectively.

Differences Among "Limit," "Tolerance," and "Allowance"

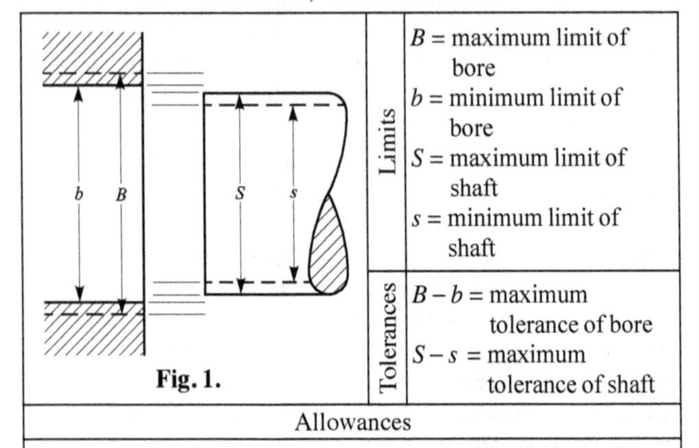

Fig. 1.

Limits	B = maximum limit of bore b = minimum limit of bore S = maximum limit of shaft s = minimum limit of shaft
Tolerances	$B - b$ = maximum tolerance of bore $S - s$ = maximum tolerance of shaft

Allowances
$B - s$ = maximum allowance, or if s is greater than B (as for tight or forced fits) then $s - B$ = minimum allowance for fit.
$b - S$ = minimum allowance, or if S is greater than b (as for tight or forced fits) then S − b = maximum allowance for fit.

The maximum pitch diameter or limit of the screw = 0.9168, and the minimum pitch diameter = 0.9100; hence, the tolerance = 0.9168 − 0.9100 = 0.0068 inch. The nut tolerance = 0.9276 − 0.9100 = 0.0176 inch. The maximum allowance for medium fit = maximum pitch diameter of nut − minimum pitch diameter of screw = 0.9276 − 0.9168 = 0.0108 inch. The minimum allowance = minimum pitch diameter of nut − maximum pitch diameter of screw = 0.9188 − 0.9168 = 0.0020.

Relation of Tolerances to Limiting Dimensions and How Basic Size Is Determined.—The absolute limits of the various dimensions and surfaces indicate danger points, in as much as parts made beyond these limits are unserviceable. A careful analysis of a mechanism shows that one of these danger points is more sharply defined than the other. For example, a certain stud must always assemble into a certain hole. If the stud is made beyond its maximum limit, it may be too large to assemble. If it is made beyond its minimum limit, it may be too loose or too weak to function. The absolute maximum limit in this case may cover a range of 0.001 inch, whereas the absolute minimum limit may have a range of at least 0.004 inch. In this case the maximum limit is the more sharply defined.

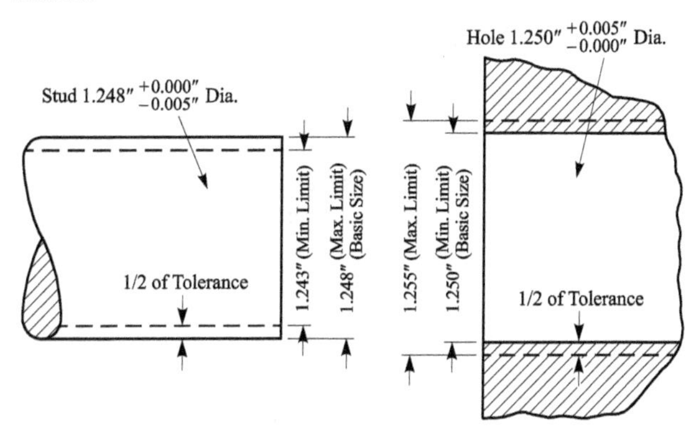

Fig. 2. Graphic Illustration of Basic Size or Dimension

The basic size expressed on the component drawing is that limit that defines the more vital of the two danger points, while the tolerance defines the other. In general, the basic dimension of a male part such as a shaft is the maximum limit that requires a minus tolerance. Similarly, the basic dimension of a female part is the minimum limit requiring a plus tolerance, as shown in **Fig. 2**. There are, however, dimensions that define neither a male nor a female surface, such as, for example, dimensions for the location of holes. In a few such instances, a variation in one direction is less dangerous than a variation in the other. Under these conditions, the basic dimension represents the danger point, and the unilateral tolerance permits a variation only in the less dangerous direction. At other times, the conditions are such that any variation from a fixed point in either direction is equally dangerous. The basic size then represents this fixed point, and tolerances on the drawing are bilateral and extend equally in both directions. (See *Handbook* page **641** for explanation of unilateral and bilateral tolerances.)

When Allowance Provides Clearance Between Mating Parts.— When one part must fit freely into another part like a shaft in its bearing, the allowance between the shaft and bearing represents a clearance space. It is evident that the amount of clearance varies widely for different classes of work. The minimum clearance should be as small as will permit the ready assembly and operation of the parts, while the maximum clearance should be as great as the functioning of the mechanism will allow. The difference between the maximum and minimum clearances defines the extent of the tolerances. In general, the difference between the basic sizes of companion parts equals the minimum clearance (see **Fig. 3**), and the term "allowance," if not defined as maximum or minimum, is quite commonly applied to the minimum clearance.

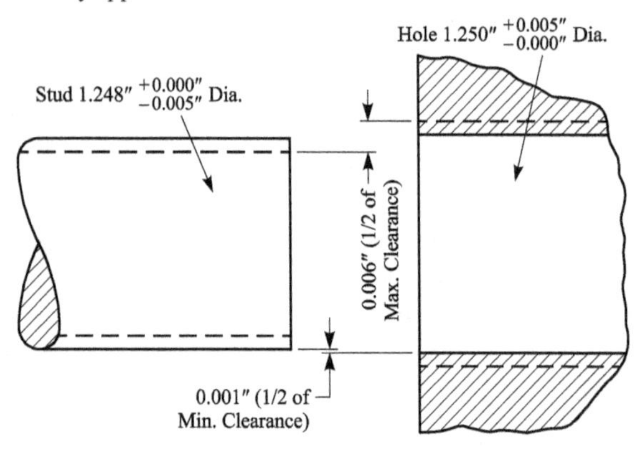

Fig. 3. Graphic Illustration of Maximum and Minimum Clearance

When Interference of Metal Is the Result of Allowance.— If a shaft or pin is larger in diameter than the hole into which it is forced, there is, of course, interference between the two parts. The metal surrounding the hole is expanded and compressed as the shaft or other part is forced into place.

Engine crankpins, car axles, and various other parts are assembled in this way (see paragraph *Allowance for Forced Fits*,

Handbook page **643**). The force and shrink fits in **Table 11** (starting on *Handbook* page **659**) all represent interference of metal.

If interchangeable parts are to be forced together, the minimum interference establishes the danger point. Thus, for force fits, the basic dimension of the shaft or pin is the minimum limit requiring a plus tolerance, and the basic dimension of the hole is the maximum limit requiring a minus tolerance, see **Fig. 4**.

Obtaining Allowance by Selection of Mating Parts.—The term "selective assembly" is applied to a method of manufacturing that is similar in many of its details to interchangeable manufacturing. In selective assembly, the mating parts are sorted according to size and assembled or interchanged with little or no further machining or hand work.

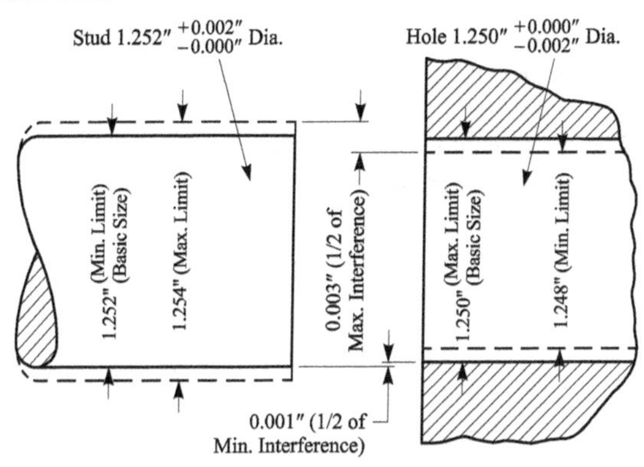

Fig. 4. Illustration of the Meaning of Maximum and Minimum Interference

The chief purpose of manufacturing by selective assembly is the production of large quantities of duplicate parts as economically as possible. As a general rule, the smaller the tolerances, the more exacting and expensive will be the manufacturing processes. However, it is possible to use comparatively large tolerances and then reduce them, in effect, by selective assembly, provided the quantity of parts is large enough to make such selective fitting possible. To illustrate, **Fig. 5** shows a plug or stud that has a plus tolerance of 0.001 inch and a hole that also has a plus tolerance of 0.001 inch. Assume that this tolerance of 0.001 inch represents the normal size variation on each part when manufactured efficiently. With this tolerance, a minimum plug in a maximum hole would have a clearance 0.2510 – 0.2498 = 0.0012 inch, and a maximum plug in a minimum hole would have a *metal interference* of 0.2508 – 0.2500 = 0.0008 inch. Suppose, however, that the clearance required for these parts must range from zero to 0.0004 inch. This reduction can be obtained by dividing both plugs and holes into five groups. (See below.) Any studs in Group *A*, for example, will assemble in any hole in Group *A*, but the studs in one group will not assemble properly in the holes in another group. When the largest stud in Group *A* is assembled in the smallest hole in Group *A*, the clearance equals zero. When the smallest stud in Group *A* is assembled in the largest hole in Group *A*, the clearance equals 0.0004 inch. Thus, in selective assembly manufacturing, there is a double set of limits, the first being the manufacturing limits and the second the assembling limits. Often, two separate drawings are made of a part that is to be graded before assembly. One shows the manufacturing tolerances only, so as not to confuse the operator, and the other gives the proper grading information.

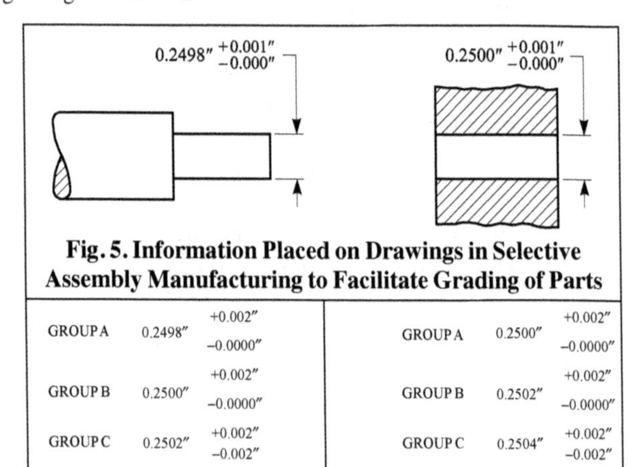

Fig. 5. Information Placed on Drawings in Selective Assembly Manufacturing to Facilitate Grading of Parts

GROUP A	0.2498″	+0.002″ −0.0000″	GROUP A	0.2500″	+0.002″ −0.0000″
GROUP B	0.2500″	+0.002″ −0.0000″	GROUP B	0.2502″	+0.002″ −0.0000″
GROUP C	0.2502″	+0.002″ −0.002″	GROUP C	0.2504″	+0.002″ −0.002″
GROUP D	0.2504″	+0.0000″ −0.002″	GROUP D	0.2506″	+0.0000″ −0.002″
GROUP E	02506″	+0.0000″ −0.002″	GROUP E	0.2508″	+0.0000″ −0.002″

Example 3: Force and shrink fit data are given in **Table 11**, page **659** of the *Handbook*. Establish the limits of size and interference of the hole and shaft for a Class FN-1 fit of 2-inch diameter.

$$\text{Max. hole} = 2 + 0.0007 = 2.0007;\ \text{min. hole} = 2 - 0 = 2$$

$$\text{Max. shaft} = 2 + 0.0018 = 2.0018;\ \text{min. shaft} = 2 + 0.0013 = 2.0013$$

In the second column of the table, the minimum and maximum interferences are given as 0.0006 and 0.0018 inch, respectively, for a FN-1 fit of 2-inch diameter. For a *selected* fit, shafts are selected that are 0.0012 inch larger than the mating holes; that is, for any mating pair, the shaft is larger than the hole by an amount midway between the minimum (0.0006-inch) and maximum (0.0018 inch) interference.

Dimensioning Drawings to Ensure Obtaining Required Tolerances.—In dimensioning the drawings of parts requiring tolerances, there are certain fundamental rules that should be applied.

Rule 1: In interchangeable manufacturing there is only one dimension (or group of dimensions) in the same straight line that can be controlled within fixed tolerances. This dimension is the distance between the cutting surface of the tool and the locating or registering surface of the part being machined. Therefore, it is incorrect to locate any point or surface with tolerances from more than one point in the same straight line.

Rule 2: Dimensions should be given between those points that it is essential to hold in a specific relation to each other. Most dimensions, however, are relatively unimportant in this respect. It is good practice to establish common location points in each plane and give, as far as possible, all such dimensions from these points.

Rule 3: Basic dimensions given on component drawings for interchangeable parts should be, except for force fits and other unusual conditions, the *maximum metal* size (maximum shaft or plug and minimum hole). The direct comparison of the basic sizes should check the danger zone, which is the minimum clearance condition in most instances. It is evident that these sizes are the most important ones, as they control interchangeability, and should be the first determined. Once established, they should remain fixed if the mechanism functions properly, and the design is unchanged.

The direction of the tolerances, then, would be that which recedes from the danger zone. In most instances, this directionality means that the direction of the tolerances is that which will increase the clearance. For force fits, basic dimensions determine the minimum interference, and tolerances limit the maximum interference.

Rule 4: Dimensions must not be duplicated between the same points. The duplication of dimensions causes much needless trouble, due to changes being made in one place and not in the others. It is easier to search a drawing to find a dimension than it is to have them duplicated and more readily found but inconsistent.

Rule 5: As far as possible, the dimensions on comparison parts should be given from the same relative locations. Such a procedure assists in detecting interferences and other improper conditions.

In attempting to work in accordance with general laws or principles, one other elementary rule should always be kept in mind. Special requirements need special consideration. The following detailed examples are given to illustrate the application of the five rules and to indicate results of their violation.

Violations of Rules for Dimensioning.—Fig. 6 shows a very common method of dimensioning a part such as the stud shown, but one that is bad practice as it violates the first and second rules. The dimensions given for the diameters are correct, so they are eliminated from the discussion. The dimensions given for the various lengths are wrong: First, because they give no indication as to the essential lengths; second, because of several possible sequences of operations, some of which would not maintain the specified conditions.

Fig. 7 shows one possible sequence of operations indicated alphabetically. If we first finish the dimension *a* and then finish *b*, the dimension *c* will be within the specified limits. However, the dimension *c* is then superfluous. **Fig. 8** gives another possible sequence of operations. If we first establish *a*, and then *b*, the dimension *c* may vary 0.030 instead of 0.010 inch as specified in **Fig. 6**. **Fig. 9** gives a third possible sequence of operations. If we first finish the overall length *a*, and then the length of the body *b*, the stem *c* may vary 0.030 inch instead of 0.010 inch as specified in **Fig. 6**.

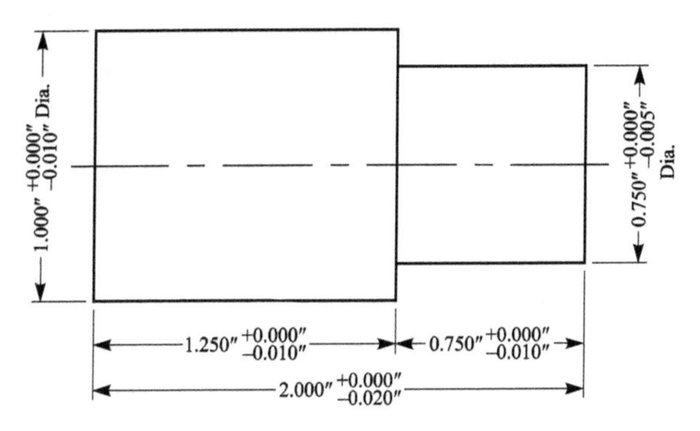

Fig. 6. Common but Incorrect Method of Dimensioning

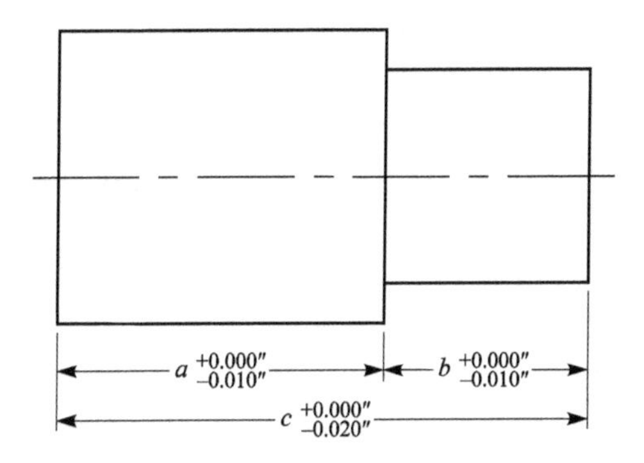

Fig. 7. One Interpretation of Dimensioning in Fig. 6

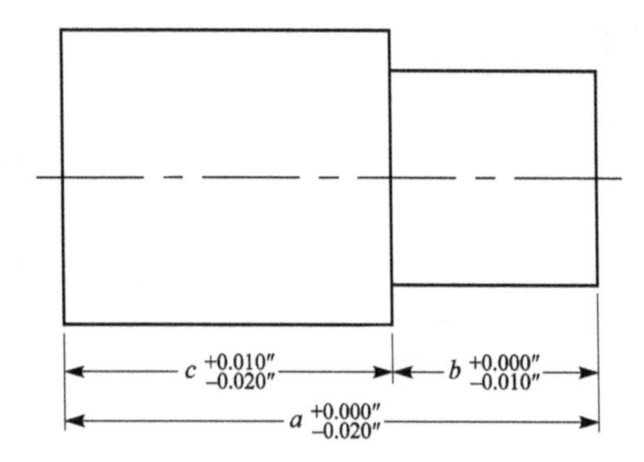

Fig. 8. Another Interpretation of Dimensioning in Fig. 6

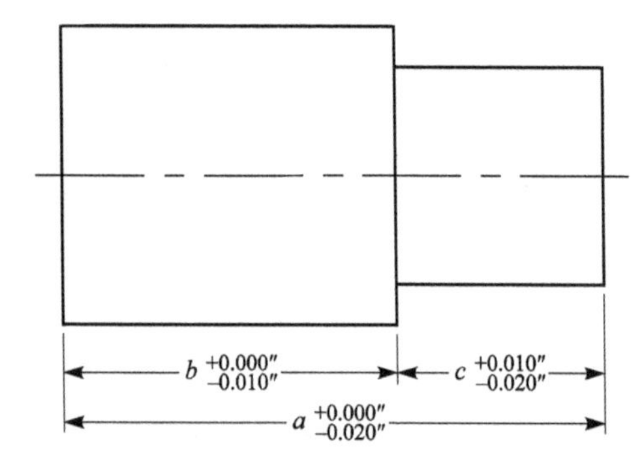

Fig. 9. A Third Interpretation of Dimensioning in Fig. 6

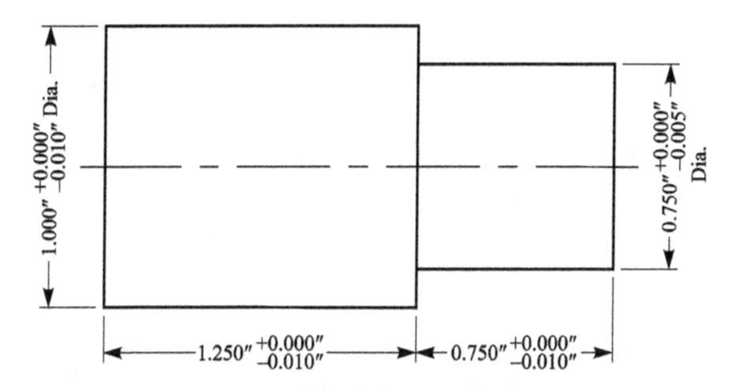

Fig. 10. Correct Dimensioning if Length of Body and Length of Stem Are Most Important

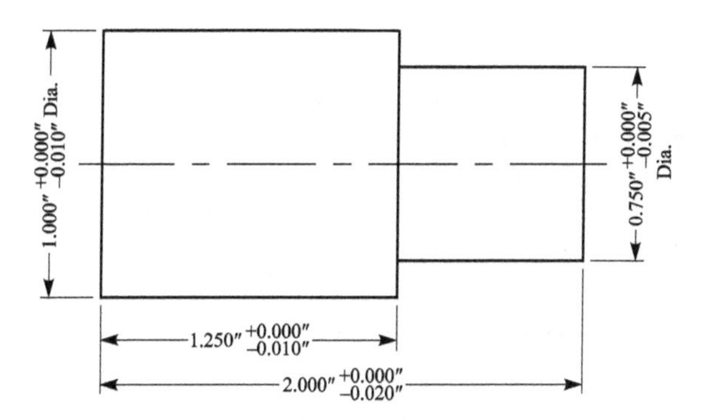

Fig. 11. Correct Dimensioning if Length of Body and Overall Length Are Most Important

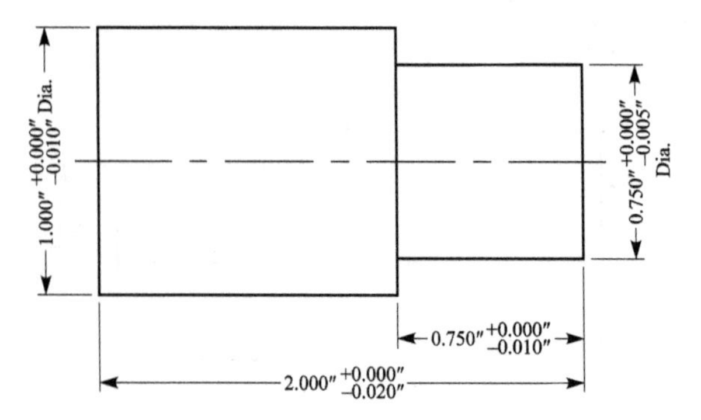

Fig. 12. Correct Dimensioning if Overall Length and Length of Stem Are Most Important

If three different plants were manufacturing this part, each one using a different sequence of operations, it is evident from the foregoing that a different product would be received from each plant. The example given is the simplest one possible. As the parts become more complex, and the number of dimensions increases, the number of different combinations possible and the extent of the variations in size that will develop also increase.

Fig. 10 shows the correct way to dimension this part if the length of the body and the length of the stem are the essential dimensions. **Fig. 11** is the correct way if the length of the body and the length overall are the most important. **Fig. 12** is correct if the length of the stem and the length overall are the most important. If the part is dimensioned in accordance with **Fig. 10**, **Fig. 11**, or **Fig. 12**, then the product from any number of factories should be alike.

PRACTICE EXERCISES FOR SECTION 12

(See *Answers to Practice Exercises for Section 12* on page **243**)

1) What factors influence the allowance for a forced fit?

2) What is the general practice in applying tolerances to center distances between holes?

3) A 2-inch shaft is to have a tolerance of 0.003 inch on the diameter. Show, by examples, three ways of expressing the shaft dimensions.

4) In what respect does a bilateral tolerance differ from a unilateral tolerance? Give an example that demonstrates this difference.

5) What is the relationship between gagemaker's tolerance and workplace tolerance?

6) Name the different classes of fits for screw thread included in the American Standards.

7) How does the Unified screw for screw threads differ from the former American Standard with regard to clearance between mating parts? With regard toward working tolerance?

8) Under what conditions is one limiting dimension or "limit" also a basic dimension?

9) According to table at the bottom of *Handbook* page **650**, broaching will produce work within tolerance grades 5 through 8. What does this mean in terms of thousandths of an inch, considering a 1-inch diameter broached hole?

10) Does surface roughness affect the ability to work within the tolerance grades specified in Exercise 10?

SECTION 13

USING STANDARDS DATA AND INFORMATION

(References to Standards appear throughout the *Machinery's Handbook, 32nd Edition*)

Standards are needed in metalworking manufacturing to establish dimensional and physical property limits for parts that are to be interchangeable. Standards make it possible for essential parts, such as nuts, screws, bolts, splines, gears, etc., to be manufactured at different times and places with the assurance that they will meet assembly requirements. Standards are also needed for commonly used tools, such as twist drills, reamers, milling cutters, etc., so that only a given number of sizes need be made available to cover a given range and to ensure adequate performance. Also, performance standards often are established to make sure that machines and equipment will satisfy their application requirements.

A specific standard may be established by a company on a limited basis for its own use. An industry may find that a standardized set of specifications is needed, and its member companies working through their trade association come to an agreement as to what requirements should be included. Sometimes, industry standards sponsored by a trade association or an engineering society become acceptable by a wide range of consumers, manufacturers, and government agencies as national standards and are made available through a national authority such as the American National Standards Institute (ANSI) and the American Society of Mechanical Engineers (ASME). More countries have been deciding that standards should be universal and are working to this end through the International Organization for Standardization (ISO) and other international bodies.

In the United States and some other English-speaking countries, there are two systems of measurement in use: the inch system and the metric system. As a result, standards for, say, bolts, nuts, and screws have been developed for both inch and metric dimensions as will be found in *Machinery's Handbook*. However, an increasing number of multinational corporations and their local suppliers are finding it prohibitively expensive to operate with two systems of measurements and standards. Thus, in order to use available expertise in one plant location, a machine may be designed in an "inch" nation only to be produced later in a "metric" country or vice versa. This situation generates additional costs in the conversion of drawings, substitution of equivalent standard steel sizes and fasteners, and conversion of testing and material specifications, etc. Because of these problems, more and more official Standards are being developed and published in the United States and throughout the world that are based, wherever practicable, upon established ISO and other authoritative organizational Standards.

In the *Handbook*, the user will find that a large number of both inch and metric Standards data and information are provided. It should be noted that at the head of each table of data derived from a specific Standard, the source is given in parentheses, such as (ANSI/ASME B18.3-2012). ANSI indicates the American National Standards Institute and ASME stands for the American Society of Mechanical Engineers; B18.3 is the identifying number of the Standard; and 2012 is the date the Standard was published, or revised, and became effective. If there is an additional year after the publishing year with an "R" in front of it, the Standard was "reaffirmed" during that year. And even when Standards are officially withdrawn, they may still be referenced for important content.

Generally, new products are produced to the metric Standards; older products and replacement parts for them may require reference to older inch Standards, and some products, such as inch-unit pipe threads, are considered as standard for the near future because of widespread use throughout the world.

Important Objectives of Standardization.—The purpose of standardization is to manufacture goods for less direct and indirect costs and to provide finished products that meet the demands of the marketplace. A more detailed description of the objectives could be as follows:

Lower the production costs when the aim is to:

1) Facilitate and systematize the work of skilled designers;

2) Ensure optimum selection of materials, components, and semi-finished products;

3) Reduce stocks of materials, semifinished products, and finished products;

4) Minimize the number of different products sold; and

5) Facilitate and reduce the cost of procurement of purchased goods.

Meet the demands of the market place, when the objective is to:

1) Conform to regulations imposed by government and trade organizations;

2) Stay within safety regulations set forth by governments; and

3) Facilitate interchangeability requirements with existing products.

Standardization Technique.—The two commonly used basic principles for the preparation of a Standard are:

1) Analytical standardization–Standard developed from scratch.

2) Conservative standardization–Standard based, so far as is possible, on existing practice.

In practice, it appears that a Standard cannot be prepared completely by one or the other of the two methods but emerges from a compromise between the two. The goal of the standardization technique, then, should be to utilize the basic material and the rules and the aids available in such a way that a valid and practical compromise solution is reached.

The basic material could consist of such items as former company Standards, vendor catalog data, national and international Standards, requirements of the company's customers, and competitor's material. Increasingly important are the national and international Standards in existence on the subject; they should always play an important part in any conservative standardization work. For example, it would be foolish to create a new metric Standard without first considering some existing European metric Standards.

Standards Information in the *Handbook*.—Among the many kinds of material and data to be found in the *Handbook*, the user will note that extensive coverage is given to Standards of several types: American National Standards, British Standards, ISO Standards, engineering society Standards, trade association Standards, and, in certain instances, company product Standards. Both inch and metric system Standards are given wherever appropriate. Inch dimension Standards sometimes are provided only for use during transition to metric Standards or to provide information for the manufacture of replacement parts.

In selecting Standards to be presented in the *Handbook*, the editors have chosen those Standards most appropriate to the needs of *Handbook* users. Text, illustrations, formulas, tables of data, and examples have been arranged in the order best suitable for direct and quick use. As an example of this type of presentation, the section on bevel gearing, starting on *Handbook* page **2258**, begins with text material that provides the basis for understanding information presented in the AGMA Standards; the illustrations on *Handbook* pages **2263–2264** provide visual definition of essential parts and dimensions of a bevel gear; the formulas on *Handbook* page **2252** show how to calculate dimensions of milled bevel gears; the tables starting on *Handbook* page **2266** give numbers of formed cutters used to mill teeth in mating bevel gear and pinion sets with shafts at right angles; and finally, the worked-out examples beginning on *Handbook* page **2268** give a step-by-step procedure for selecting formed cutters for milling bevel gears. Also, where combinations of tables and formulas are given, the formulas have been arranged in the best sequence for computation with the aid of a pocket calculator.

"Soft" Conversion of Inch to Metric Dimensions.—The dimensions of certain products, when specified in inches, may be converted to metric dimensions, or vice versa, by multiplying by the appropriate conversion factor so that the parts can be fabricated either to inch or to the equivalent metric dimensions and still be fully interchangeable. Such a conversion is called a *soft* conversion. An example of a soft conversion is available on *Handbook* page **2476**, which gives the inch dimensions of standard lockwashers for ball bearings. The footnote to the table indicates that multiplication of the tabulated inch dimensions by 25.4 and rounding the results to two decimal places will provide the equivalent metric dimensions.

"Hard" Metric or Inch Standard Systems.—In a *hard* system, those dimensions in the system that have been standardized cannot be converted to another dimensional system that has been standardized independently of the first system. As stated in the footnote on page **2353** of the *Handbook*, "In a 'hard' system the tools of production, such as hobs, do not bear a usable relation to the tools in another system; i.e., a 10 diametral pitch hob calculates to be equal to a 2.54 module hob in the metric module system, a hob that does not exist in the metric standard."

Interchangeability of Parts Made to Revised Standards.—Where a Standard has been revised, there may still remain some degree of interchangeability between older parts and those made to the new Standard. As an example, starting on page **2344** of the *Handbook*, there are two tables showing which of the internal and external involute splines made to older Standards will mate with those made to newer Standards.

PRACTICE EXERCISES FOR SECTION 13

(See *Answers to Practice Exercises for Section 13* on page **244**)

1) What is the breaking strength of a 6 × 7 fiber-core wire rope ¼ inch in diameter if the rope material is mild plow steel?

2) What factor of safety should be applied to the rope in Exercise 1?

3) How many carbon steel balls of ¼-inch diameter would weigh 1 lb? How would this information be obtained without the table?

4) For a 1-inch diameter of shaft, what size square key is appropriate?

5) Find the hole size needed for a 5⁄32-inch standard cotter pin.

6) Find the limits of size for a 0.1250-inch diameter hardened and ground dowel pin.

7) For a 3AM1-17 retaining ring (snap ring), what is the maximum allowable speed of rotation?

8) Find the hole size required for a type AB steel thread-forming screw of number 6 size in 0.105-inch-thick stainless steel.

SECTION 14

STANDARD SCREW AND PIPE THREADS

Machinery's Handbook pages **1943–2150**

Different screw thread forms and standards have been originated and adopted at various times, either because they were considered superior to other forms or because of the special requirements of screws used for a certain class of work.

A standard thread conforms to an adopted Standard with regard to the form or contour of the thread itself and as to the pitches or numbers of threads per inch for different screw diameters.

The United States Standard formerly used in the United States was replaced by an American Standard having the same thread form as the former Standard and a more extensive series of pitches, as well as tolerances and allowances for different classes of fits. This American Standard was revised in 1949 to include a Unified Thread Series, which was established to obtain screw thread interchangeability among the United Kingdom, Canada, and the United States.

The Standard was revised again in 1959. The Unified threads are now generally used in the United States and the former American Standard threads are now used only in certain applications where the changeover in tools, gages, and manufacturing has not been completed. The differences between Unified and the former National Standard threads are explained on pages **1943** and **1950** in the *Handbook*.

As may be seen in the table on *Handbook* page **1953**, the Unified series of screw threads consists of three Standard series having graded pitches (UNC, UNF, and UNEF) and eight Standard series of uniform (constant) pitch. In addition to these Standard series. There are places in the table beginning on *Handbook* page **1954** where special threads (UNS) are listed. These UNS threads are for use only if Standard series threads do not meet requirements.

Example 1: The table on *Handbook* page **1981** shows that the pitch diameter of a 2-inch screw thread is 1.8557 inches. What is meant by the term "pitch diameter" as applied to a screw thread and how is it determined?

According to a definition given in connection with American Standard screw threads, the *pitch diameter* of a straight (nontapering) screw thread is the diameter of an imaginary cylinder, the surface of which would pass through the threads at such points as to make equal the width of the threads and the width of the spaces cut by the surface of the cylinder.

The basic pitch diameter equals the basic major (outside) diameter minus two times the addendum of the external thread (*Handbook* page **1952**), so the basic pitch diameter for the 2-inch example, with $4\frac{1}{2}$ threads per inch, is $2.00 - 2 \times 0.07217 = 1.8557$ inches.

Example 2: The tensile strength of a bolt $3\frac{1}{2}$ inches in diameter at a stress of 6000 pounds per square inch may be calculated by means of the formulas on *Handbook* page **1668**. This formula uses the largest diameter of the bolt, avoiding the need to take account of the reduced diameter at the thread root, and gives a tensile strength of 35,175 pounds for the conditions noted.

If the second formula on page **1668**, based on the area of the smallest diameter, is used for the same bolt and stress, and the diameter of the thread root is taken as 3.1 inches, then the tensile strength is calculated as 40,636 pounds. The difference in these formulas is that the first uses a slightly greater factor of safety than the second, taking account of possible variations in thread depth.

Example 3: Handbook page **2131** gives formulas for checking the pitch diameter of screw threads by the three-wire method (when effect of lead angle is ignored). Show how these formulas have been derived using the one for the American National Standard Unified thread as an example.

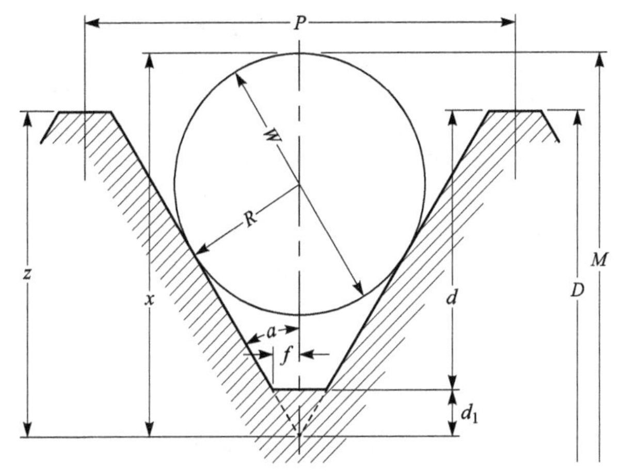

Fig. 1. Diagram Illustrating the Derivation of Formulas for Three-Wire Measurements of Screw Thread Pitch Diameters

It is evident from **Fig. 1** that:

$$M = D - 2z + 2x \tag{1}$$

$$x = R + \frac{R}{\sin a} \text{ and } 2x = 2R + \frac{2R}{0.5}\text{; hence,}$$

$$2x = \frac{(2 \times 0.5 + 2)R}{0.5} = \frac{3R}{0.5} = 6R = 3W$$

$$z = d + d_1 = 0.6495P + f \times \cot\alpha$$

$$f = 0.0625P\text{; therefore,}$$

$$z = 0.6495P + 0.10825P = 0.75775P$$

If, in **Formula (1)**, we substitute the value of $2z$ or $2 \times 0.75775P$ and the value of $2x$, we have:

$$M = D - 1.5155 \times P + 3W \tag{2}$$

This **Formula (2)** is the one found in previous editions of the *Handbook*. In the 22nd and subsequent editions of the *Handbook* use of the outside diameter D in **Formula (2)** above was eliminated to provide a formula in terms of the pitch diameter E. Such a formula is useful for finding the wire measurement corresponding to the actual pitch diameter, whether it be correct, undersize, or oversize.

According to the last paragraph of **Example 1**, above, $E = D - 2 \times$ thread addendum. On *Handbook* page **1952**, the formula for thread addendum given at the top of the last column is $0.32476P$. Therefore, $E = D - 2 \times 0.32476P$, or, transposing this formula, $D = E + 2 \times 0.32476P = E + 0.64952P$. Substituting this value of D into **Formula (2)** gives: $M = E + 0.64952P - 1.5155P + 3W = E - 0.8660P + 3W$, which is the current *Handbook* formula.

Example 4: On *Handbook* page **2138**, a formula is given for checking the angle of a screw thread by a three-wire method. How is this formula derived? By referring to **Fig. 2**,

$$\sin a = \frac{W}{S} \tag{1}$$

If D = diameter of larger wires and d = diameter of smaller wires,

$$W = \frac{D - d}{2}$$

If B = difference in measurement over wires, then the difference S between the centers of the wires is:

$$S = \frac{B - (D - d)}{2}$$

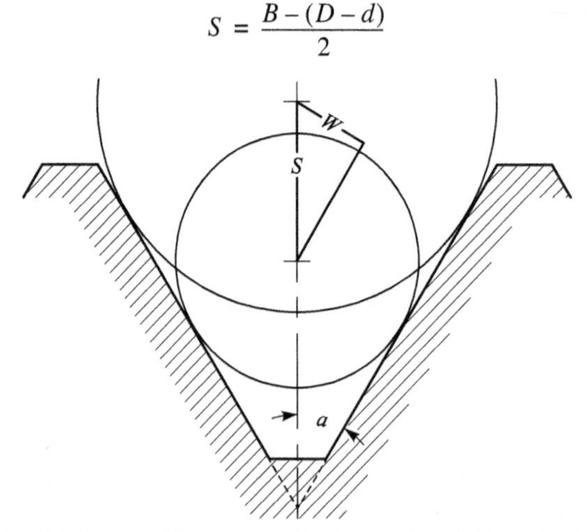

Fig. 2. Diagram Illustrating the Derivation of Formula for Checking the Thread Angle by the Three-Wire System

By inserting these expressions for W and S in **Formula (1)** and canceling, the formula given in the *Handbook* is obtained if A is substituted for $D-d$.

$$\sin a = \frac{A}{B-A}$$

Example 5: A vernier gear-tooth caliper (like the one shown on *Handbook* page **2229**) is to be used for checking the width of an Acme screw by measuring squarely across or perpendicular to the thread. Since standard screw thread dimensions are in the plane of the axis, how is the width square or normal to the sides of the thread determined? Assume that the width is to be measured at the pitch line and that the number of threads per inch is two.

The table on *Handbook* page **2060** shows that for two threads per inch, the depth is 0.260 inch; hence, if the measurement is to be at the pitch line, the vertical scale of the caliper is set to $(0.260-0.010)\div 2 = 0.125$ inch. The pitch equals

$$\frac{1}{\text{No. of threads per inch}} = \tfrac{1}{2} \text{ inch}$$

The width A, **Fig. 3**, in the plane of the axis equals $\tfrac{1}{2}$ the pitch, or $\tfrac{1}{4}$ inch. The width B perpendicular to the sides of the thread = width in axial plane × cosine helix angle.

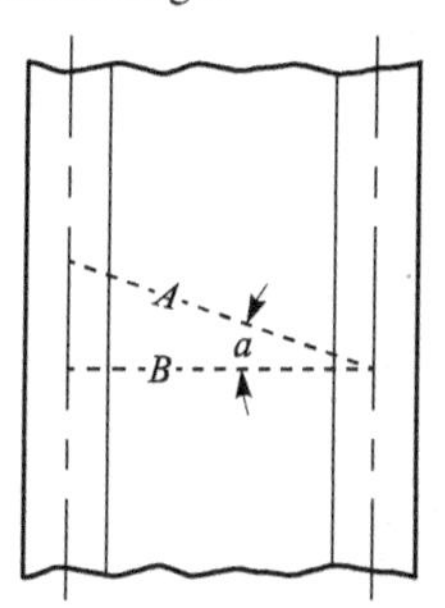

Fig. 3. Determining the Width Perpendicular to the Sides of a Thread at the Pitch Line

(The helix angle, which equals angle a, is based upon the pitch diameter and is measured from a plane perpendicular to the axis of the screw thread.) The width A in the plane of the axis represents the hypotenuse of a right triangle, and the required width B equals the side adjacent; hence width $B = A \times$ cosine of helix angle. The angle of the thread itself (29° for an Acme Thread) does not affect the solution.

Width of Flat End of Unified Screw Thread and American Standard Acme Screw Thread Tools.—The widths of the flat or end of the threading tool for either of these threads may be measured by using a micrometer as illustrated at **(A)** in **Fig. 4**. In measuring the thread tool, a scale is held against the spindle and anvil of the micrometer, and the end of the tool is placed against this scale. The micrometer is then adjusted to the position shown and 0.2887 inch subtracted from the reading for an American Standard screw-thread tool. For American Standard Acme threads, 0.1293 inch is subtracted from the micrometer reading to obtain the width of the tool point. The constants (0.2887 and 0.1293), which are subtracted from the micrometer reading, are only correct when the micrometer spindle has the usual diameter of 0.25 inch.

An ordinary gear-tooth vernier caliper also may be used for testing the width of a thread tool point, as illustrated at **(B)** in **Fig. 4**. If the measurement is made at a vertical distance x of $\frac{1}{4}$ inch from the points of the caliper jaws, the constants previously given for American Standard caliper reading to obtain the actual width of the cutting end of the tool.

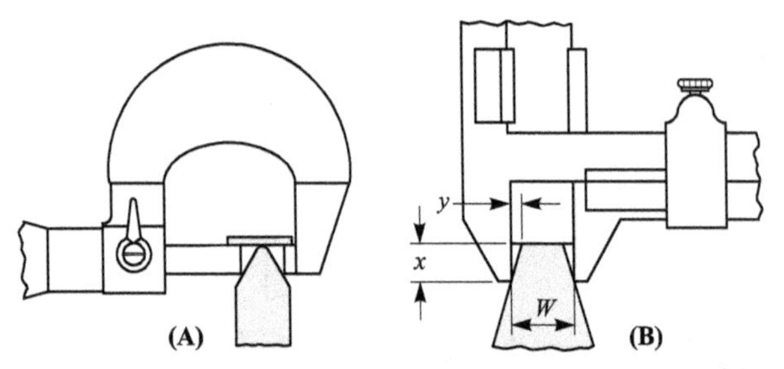

Fig. 4. Measuring Width of Flat on Threading Tool (A) with a Micrometer; (B) with a Gear-Tooth Vernier

Example 6: Explain how the constants 0.2887 and 0.1293 referred to in a preceding paragraph are derived and deduce a general rule applicable regardless of the micrometer spindle diameter or vertical dimension x, **Fig. 4**.

The dimension x (which also is equivalent to the micrometer spindle diameter) represents one side of a right triangle (the side adjacent), having an angle of $29 \div 2 = 14$ degrees and 30 minutes, in the case of an Acme thread. The side opposite or y = side adjacent × tangent = dimension $x \times \tan 14° 30'$.

If x equals 0.25 inch, then side opposite or $y = 0.25 \times 0.25862 = 0.06465$; hence, the caliper reading minus 2×0.06465 = width of the flat end ($2 \times 0.06465 = 0.1293$ = constant).

The same result would be obtained by multiplying 0.25862 by $2x$; hence, the following rule: To determine the width of the end of the threading tool, by the general method illustrated in **Fig. 4**, multiply twice the dimension x (or spindle diameter in the case of the micrometer) by the tangent of one-half the thread-tool angle, and subtract this product from the width w to obtain the width at the end of the tool.

Example 7: A gear-tooth vernier caliper is to be used for measuring the width of the flat of an American Standard external screw-thread tool. The vertical scale is set to $\frac{1}{8}$ inch (corresponding to the dimension x, **Fig. 4**). How much is subtracted from the reading on the horizontal scale to obtain the width of the flat end of the tool?

$$\tfrac{1}{8} \times 2 \times \tan 30° = \tfrac{1}{4} \times 0.57735 = 0.1443 \text{ inch}$$

Hence, the width of the flat equals w, **Fig. 4**, minus 0.1443. This width should be equal to one-eighth of the pitch of the thread to be cut, since this is the width of flat at the minimum minor diameter of American Standard external screw threads.

PRACTICE EXERCISES FOR SECTION 14

(See *Answers to Practice Exercises for Section 14* on page **244**)

1) What form of screw thread is most commonly used (*a*) in the United States? (*b*) in Britain?

2) What is the meaning of abbreviations 3″-4NC-2?

3) What are the advantages of an Acme thread compared to a square thread?

4) For what reason would a Stub Acme thread be preferred in some applications?

5) Find the pitch diameters of the following screw threads of American Standard Unified form: $\frac{1}{4}$ – 28 (meaning $\frac{1}{4}$-inch diameter and 28 threads per inch); $\frac{3}{4}$ – 10?

6) How much taper is used on a standard pipe thread?

7) Under what conditions are straight, or nontapering, pipe threads used?

8) In cutting a taper thread, what is the proper position for the lathe tool?

9) If a lathe is used for cutting a British Standard pipe thread, in what position is the tool set?

10) A thread tool is to be ground for cutting an Acme thread having 4 threads per inch; what is the correct width of the tool at the end?

11) What are the common shop and toolroom methods of checking the pitch diameters of American Standard screw threads requiring accuracy?

12) In using the formula on *Handbook* page **2131** for measuring an American Standard screw thread by the three-wire method, why should the constant 0.86603 be multiplied by the pitch before adding it to measurement M, even if not enclosed by parentheses?

13) What is the difference between the pitch and the lead (*a*) of a double thread? (*b*) of a triple thread?

14) In using a lathe to cut American Standard Unified threads, what should be the truncations of the tool points and the thread depths for the following pitches: 0.1, 0.125, 0.2, and 0.25 inch?

15) In using the three-wire method of measuring a screw thread, what is the micrometer reading for a $\frac{3}{4}$ - 12 special thread of American Standard form if the wires have a diameter of 0.070 inch?

16) Are most nuts made to the United States Standard dimensions?

17) Is there, at the present time, a manufacturing Standard for bolts and nuts?

18) The American Standard for machine screws includes a coarse-thread series and a fine-thread series as shown by the tables starting on *Handbook* page **1981**. Which series is commonly used?

19) How is the length (*a*) of a flat head or countersunk type of machine screw measured? (*b*) of a fillister head machine screw?

20) What size tap drill should be used for an American Standard machine screw of No. 10 size, 24 threads per inch?

21) What is the diameter of a No. 10 drill?

22) Is a No. 6 drill larger than a No. 16?

23) What is the relation between the letter size drills and the numbered sizes?

24) Why is it common practice to use tap drills that leave about $^3/_4$ of the full thread depth after tapping, as shown by the tables starting on page **2170** in the *Handbook*?

25) What form of a screw thread is used on (*a*) machine screws? (*b*) cap screws?

26) What Standard governs the pitches of cap screw threads?

27) What form of thread is used on the National Standard fire hose couplings? How many standards diameters are there?

28) In what way do hand taps differ from machine screw taps?

29) What are tapper taps?

30) The diameter of a $^3/_4$ – 10 American Standard Thread is to be checked by the three-wire method. What is the largest size wire that can be used?

31) Why is the advance of some threading dies positively controlled by a lead screw instead of relying upon the die to lead itself?

32) What is the included angle of the heads of American Standard (*a*) flat head machine screws? (*b*) flat head cap screws? (*c*) flat head wood screws?

SECTION 15

PROBLEMS IN MECHANICS

Machinery's Handbook pages **151–172**

In the design of machines or other mechanical devices, it is often necessary to deal with the actions of forces and their effects. For example, the problem may be to determine what force is equivalent to two or more forces acting in the same plane but in different directions. Another type of problem is to determine the change in the magnitude of a force resulting from the application of mechanical appliances, such as levers, pulleys, and screws used either separately or in combination. It also may be necessary to determine the magnitude of a force needed to proportion machine parts to resist the force safely; or, possibly, to ascertain if the force is enough to perform a given amount of work. Determining the amount of energy stored in a moving body or its capacity to perform work, and the power developed by mechanical apparatus, or the rate at which work is performed, are additional examples of problems frequently encountered in developing mechanical appliances. The section in *Machinery's Handbook* on mechanics, beginning on page **151**, deals with fundamental principles and formulas applicable to various mechanical problems.

The Moment of a Force.—The tendency of a force acting upon a body is, in general, to produce either a motion of translation (that is, to cause every part of the body to move in a straight line) or to produce a motion of rotation. A moment, in mechanics, is the measure of the turning effect of a force that tends to produce rotation. For example, suppose a force acts upon a body that is supported by a pivot. Unless the line of action of the force happens to pass through the pivot, the body will tend to rotate. Its tendency to rotate, moreover, will depend upon two things: (1) the magnitude of the force acting, and (2) the distance of the force from the pivot, *measuring along a line at right angles to the line of action of the force*. (See **Fig. 10** and the accompanying text on *Handbook* page **156**.)

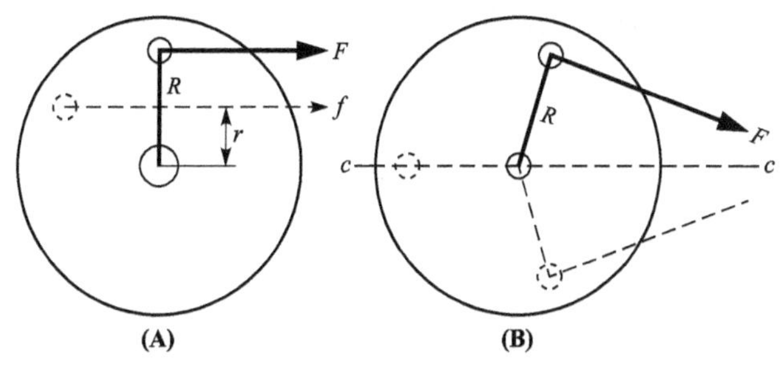

Fig. 1. Example of the Turning Moment of a Crank Disk from Zero to Maximum

Example 1: A force F of 300 pounds is applied to a crank disk (diagram A, **Fig. 1**) in the direction of the arrow. If the radius $R =$ 5 inches, what is the turning moment? Also, determine how much the turning moment is reduced when the crankpin is in the position shown by the dashed lines, assuming that the force is along line f and that $r = 2\frac{1}{2}$ inches.

When the crankpin is in the position illustrated by the solid lines, the maximum turning moment is obtained, and it equals $F \times R = 300 \times 5 = 1500$ inch-pounds or pound-inches. When the crankpin is in the position shown by the dashed lines, the turning moment is reduced one-half and equals $f \times r = 300 \times 2\frac{1}{2} = 750$ inch-pounds.

Note: Foot-pound is the unit for measurement of work and is in common use in horsepower calculations. However, torque, or turning moment, is also a unit of measurement of work. To differentiate between these two similar terms, which have the same essential meaning, it is convenient to express torque in terms of *pound-feet* (or *pound-inches*). This reversal of word sequence will readily indicate the different meanings of the two terms for units of measurement—the unit of horsepower and the unit of turning moment. A strong reason for expressing the unit of turning moment as *pound-inches* (rather than as *foot-pounds*) is because the dimensions of shafts and other machine parts ordinarily are stated in inches.

Example 2: Assume that the force F (diagram B, **Fig. 1**) is applied to the crank through a rod connecting with a crosshead that slides along center line c-c. If the crank radius $R = 5$ inches, what will be the maximum and minimum turning moments?

The maximum turning moment occurs when the radial line R is perpendicular to the force line F and equals in inch-pounds, $F \times 5$ in this example. When the radial line R is in line with the center line c-c, the turning moment is 0, because $F \times 0 = 0$. This is the *deadcenter* position for steam engines and explains why the crankpins on each side of a locomotive are located 90 degrees apart, or, in such a position that the maximum turning moment, approximately, occurs when the turning moment is zero on the opposite side. With this arrangement, it is always possible to start the locomotive since only one side at a time can be in the dead-center position.

The Principle of Moments in Mechanics.— When two or more forces act upon a rigid body and tend to turn it about an axis, then, for equilibrium to exist, the sum of the moments of the forces that tend to turn the body in one direction must be equal to the sum of the moments of those that tend to turn it in the opposite direction about the same axis.

Example 3: In **Fig. 2**, a lever 30 inches long is pivoted at the fulcrum F. At the right, and 10 inches from F, is a weight, B, of 12 pounds tending to turn the bar in a right-hand direction about its fulcrum F. At the left end, 12 inches from F, the weight A, of 4 pounds tends to turn the bar in a left-hand direction, while weight C, at the other end, 18 inches from F, has a like effect, through the use of the string and pulley P.

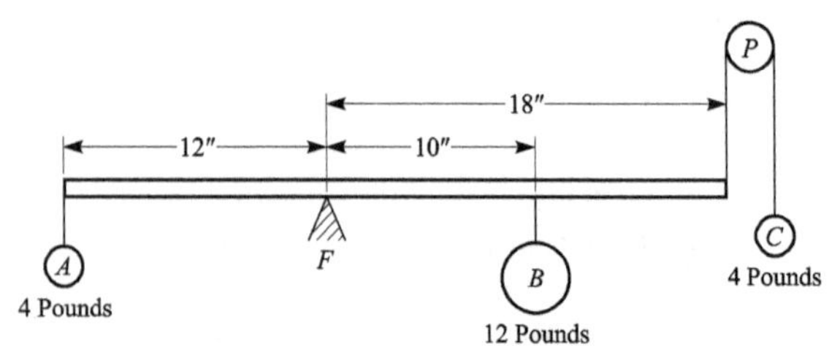

Fig. 2. Lever in Equilibrium Because the Turning Moment of a Crank Disk Varies from Zero to Maximum

Taking moments about F, which is the center of rotation, we have:

$$\text{Moment of } B = 10 \times 12 = 120 \text{ inch-pounds}$$

Opposed to this are the moments of A and C:

Moment of A	$= 4 \times 12 =$	48 inch-pounds
Moment of C	$= 4 \times 18 =$	72 inch-pounds
Sum of negative numbers	$=$	120 inch-pounds

Hence, the moments are equal, and, if we suppose, for simplicity, that the lever is weightless, it will balance or be in equilibrium. Should weight A be increased, the negative moments would be greater, and the lever would turn to the left, while if B should be increased or its distance from F be made greater, the lever would turn to the right. (See *Handbook* **Fig. 10** and the accompanying text on page **156**.)

Example 4: Another application of the principle of moments is given in **Fig. 3**. A beam of uniform cross section, weighing 200 pounds, rests upon two supports, R and S, that are 12 feet apart. The weight of the beam is considered to be concentrated at its center of gravity G, at a distance of 6 feet from each support. Each of the supports reacts or pushes upward, with a combined force equal to the downward pressure of the beam.

To make this clear, suppose two people take hold of the beam, one at each end, and that the supports are withdrawn. Then, in order to hold the beam in position, the two people must together lift or pull upward an amount equal to the weight of the beam and its load, or 250 pounds. Placing the supports in position again, and resting the beam upon them, does not change the conditions. The weight of the beam acts downward, and the supports react by an equal amount.

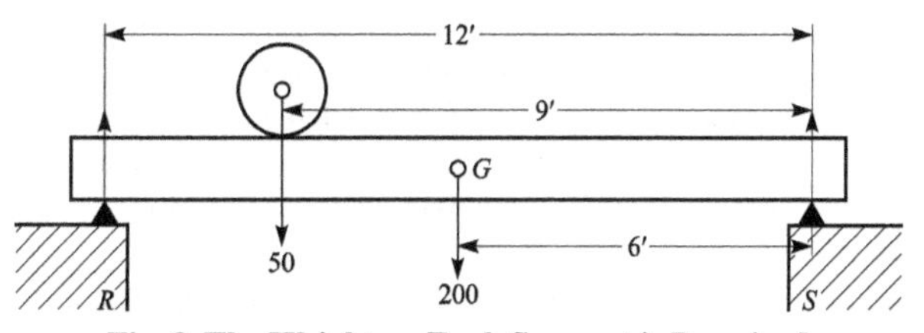

Fig. 3. The Weight on Each Support is Required

Now, to solve the problem, assume the beam to be pivoted at one support, say, at S. The forces or weights of 50 pounds and 200 pounds tend to rotate the beam in a left-hand direction about this point, while the reaction of R in an upward direction tends to give it a right-hand rotation. As the beam is balanced and has no tendency to rotate, it is in equilibrium, and the opposing moments of these forces must balance; hence, taking moments,

$$\begin{aligned} 9 \times 50 &= 450 \text{ pound-feet} \\ 6 \times 200 &= \underline{1200 \text{ pound-feet}} \\ \text{Sum of negative numbers} &= 1650 \text{ pound-feet} \end{aligned}$$

By letting R represent the reaction of support,

$$\text{Moment of } R = R \times 12 = \text{pound-feet}$$

By the principle of moments, $R \times 12 = 1650$. That is, if R, the quantity that we wish to obtain, is multiplied by 12, the result will be 1650; hence, to obtain R, divide 1650 by 12. Therefore, $R = 137.5$ pounds, which is also the weight of that end of the beam. As the total load is 250 pounds, the weight at the other end must be $250 - 137.5 = 112.5$ pounds.

The Principle of Work in Mechanics.—Another principle of more importance than the principle of moments, even in the study of machine elements, is the principle of work. According to this principle (neglecting frictional or other losses), the applied force, multiplied by the distance through which it moves, equals the resistance overcome, multiplied by the distance through which it is overcome. The principle of work may also be stated as follows:

$$\text{Work put in} = \text{lost work} + \text{work done by machine}$$

This principle holds absolutely in every case. It applies equally to a simple lever, the most complex mechanism, or to a so-called "perpetual motion" machine. No machine can be made to perform work unless a somewhat greater amount—enough to make up for the losses—is applied by some external agent. In the perpetual motion machine no such outside force is supposed to be applied, hence such a machine is impossible, and against all the laws of mechanics.

Example 5: Assume that a rope exerts a pull F of 500 pounds (upper diagram, *Handbook* page **171**) and that the pulley radius

$R = 10$ inches and the drum radius $r = 5$ inches. How much weight W can be lifted (ignoring frictional losses) and upon what mechanical principle is the solution based?

According to one of the formulas accompanying the diagram at the top of *Handbook* page **171**,

$$W = \frac{F \times R}{r} = \frac{500 \times 10}{5} = 1000 \text{ pounds}$$

This formula (and the others for finding the values of F, R, etc.) agrees with the principle of moments, and with the principle of work. The principle of moments will be applied first.

The moment of the force F about the center of the pulley, which corresponds to the fulcrum of a lever, is F multiplied by the perpendicular distance R, it being a principle of geometry that a radius is perpendicular to a line drawn tangent to a circle, at the point of tangency. Also, the opposing moment of W is $W \times r$. Hence, by the principle of moments,

$$F \times R = W \times r$$

Now, for comparison, we will apply the principle of work. Assuming this principle to be true, force F multiplied by the distance traversed by this force or by a given point on the rim of the large pulley should equal the resistance W multiplied by the distance that the load is raised. In one revolution, force F passes through a distance equal to the circumference of the pulley, which is equal to $2 \times 3.1416 \times R = 6.2832 \times R$, and the hoisting rope passes through a distance equal to $2 \times 3.1416 \times r$. Hence, by the principle of work,

$$6.2832 \times F \times R = 6.2832 \times W \times r$$

The statement simply shows that $F \times R$ multiplied by 6.2832 equals $W \times r$ multiplied by the same number, and it is evident therefore, that the equality will not be altered by canceling the 6.2832 and writing:

$$F \times R = W \times r$$

However, this statement is the same as that obtained by applying the principle of moments; hence, we see that the principle of moments and the principle of work are in harmony.

The basis of operation of a train of wheels is a continuation of the principle of work. For example, in the gear train represented by the diagram at the bottom of *Handbook* page **171**, the continued product of the applied force F and the radii of the driven wheels equals the continued product of the resistance W and the radii of the drivers. In calculations, the pitch diameters or the numbers of teeth in gear wheels may be used instead of the radii.

Efficiency of a Machine or Mechanism.—The efficiency of a machine is the ratio of the power delivered by the machine to the power received by it. For example, the efficiency of an electric motor is the ratio between the power delivered by the motor to the machinery it drives and the power it receives from the generator. Assume, for example, that a motor receives 50 kilowatts from the generator, but that the output of the motor is only 47 kilowatts. Then, the efficiency of the motor is 47 ÷ 50 = 94 percent. The efficiency of a machine tool is the ratio of the power consumed at the cutting tool to the power delivered by the driving belt. The efficiency of gearing is the ratio between the power obtained from the driven shaft to the power used by the driving shaft. Generally speaking, the efficiency of any machine or mechanism is the ratio of the output of power to the input. The percentage of power representing the difference between the input and output has been dissipated through frictional and other mechanical losses.

Mechanical Efficiency: If E represents the energy that a machine transforms into useful work or delivers at the driven end, and L equals the energy loss through friction or dissipated in other ways, then,

$$\text{Mechanical efficiency} = \frac{E}{E+L}$$

In this equation, the total energy $F + L$ is assumed to be the amount of energy that is transformed into useful and useless work. The actual total amount of energy, however, may be considerably larger than the amount represented by $E + L$. For example, in a steam engine, there are heat losses due to radiation and steam condensation, and considerable heat energy supplied to an internal combustion engine is dissipated either through the cooling water or direct to the atmosphere. In other classes of mechanical and electrical machinery, the total energy is much larger than that represented by the amount transformed into useful and useless work.

Absolute Efficiency: If E_1 equals the full amount of energy or the true total, then,

$$\text{Absolute efficiency} = \frac{E}{E_1}$$

It is evident that absolute efficiency of a prime mover, such as a steam or gas engine, will be much lower than the mechanical efficiency. Ordinarily, the term *efficiency* as applied to engines and other classes of machinery means the mechanical efficiency. The *mechanical efficiency* of reciprocating steam engines may vary from 85 to 95 percent, but the *thermal efficiency* may range from 5 to 25 percent, the smaller figure representing noncondensing engines of the cheaper class and the higher figure the best types.

Example 6: Assume that a motor driving through a compound train of gearing (see **Fig. 4**) is to lift a weight W of 1000 pounds. The pitch radius $R = 6$ inches; $R_1 = 8$ inches; pitch radius of pinion $r = 2$ inches; and radius of winding drum $r_1 = 2\frac{1}{2}$ inches. What motor horsepower will be required if the frictional loss in the gear train and bearings is assumed to be 10 percent? The pitch-line velocity of the motor pinion M is 1200 feet per minute.

The problem is to determine first the tangential force F required at the pitch line of the motor pinion; then, the equivalent horsepower is easily found. According to the formula at the bottom of *Handbook* page **171**, which does not take into account frictional losses,

$$F = \frac{1000 \times 2 \times 2\frac{1}{2}}{6 \times 8} = 104 \text{ pounds}$$

The pitch-line velocity of the motor pinion is 1200 feet per minute and, as the friction loss is assumed to be 10 percent, the mechanical efficiency equals $90 \div (90 + 10) = 0.90$ or 90 percent as commonly written; thus,

$$\text{Horsepower} = \frac{104 \times 1200}{33{,}000 \times 0.90} = 4\frac{1}{4} \text{ approximately}$$

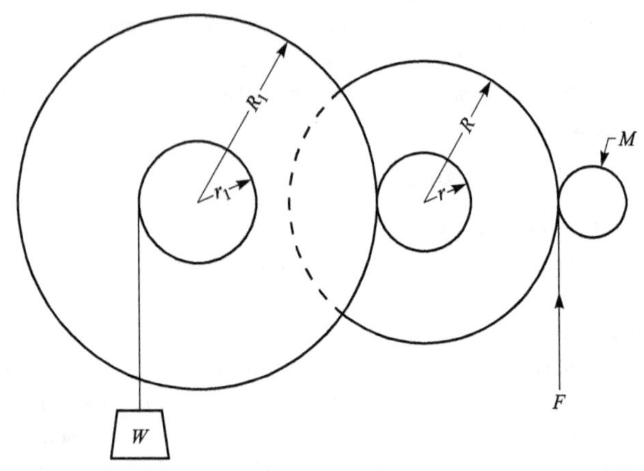

Fig. 4. Determining the Power Required for Lifting a Weight by Means of a Motor and a Compound Train of Gearing

Example 7: In designing practice, a motor of horsepower, or larger, might be selected for the drive referred to in **Example 6** (depending upon conditions) to provide extra power should it be needed. However, to illustrate the procedure, assume that the gear train is to be modified so that the calculated horsepower will be 4 instead of $4\frac{1}{4}$; conditions otherwise are the same as in **Example 6**.

$$F = \frac{33{,}000 \times 4}{1200} = 110 \text{ pounds}$$

Hence, since $W = 1000$ pounds,

$$1000 = \frac{110 \times 0.90 \times R \times R_1}{r \times r_1}$$

Insert any values for the pitch radii R, R_1, etc., that will balance the equation, so that the right-hand side equals 1000, at least approximately. Several trial solutions may be necessary to obtain a total of about 1000, at the same time, secure properly proportional gears that meet other requirements of the design. Suppose the same radii are used here, except R_1, which is increased from 8 to $8\frac{1}{2}$ inches.

Then,

$$\frac{110 \times 0.90 \times 6 \times 8\frac{1}{2}}{2 \times 2\frac{1}{2}} = 1000 \text{ approximately}$$

This example shows that the increase in the radius of the last driven gear from 8 to $8\frac{1}{2}$ inches makes it possible to use the 4-horsepower motor. The hoisting speed has been decreased somewhat, and the center distance between the gears has been increased. These changes might or might not be objectionable in actual designing practice, depending upon the particular requirements.

Force Required to Turn a Screw Used for Elevating or Lowering Loads.—In determining the force that must be applied at the end of a given lever arm in order to turn a screw (or nut surrounding it), there are two conditions to be considered: (1) when rotation is such that the load *resists* the movement of the screw, as in raising a load with a screw jack; (2) when rotation is such that the load *assists* the movement of the screw, as in lowering a load. The formulas at the bottom of the table on *Handbook* page **172** apply to both these conditions. When the load resists the screw movement, use the formula "for motion in a direction opposite to Q." When the load assists the screw movement, use the formula "for motion in the same direction as Q."

If the lead of the thread is large in proportion to the diameter so that the helix angle is large, the force F may have a negative value, which indicates that the screw will turn due to the load alone, unless resisted by a force that is great enough to prevent rotation of a nonlocking screw.

Example 8: A screw is to be used for elevating a load Q of 6000 pounds. The pitch diameter is 4 inches, the lead is 0.75 inch, and the coefficient of friction μ between screw and nut is assumed to be 0.150. What force F will be required at the end of a lever arm R of 10 inches? In this example, the load is in the direction opposite to the arrow Q (see the applicable diagram on *Handbook* page **172**).

$$F = 6000 \times \frac{0.75 + 6.2832 \times 0.150 \times 2}{6.2832 \times 2 - 0.150 \times 0.75} \times \frac{2}{10}$$
$$= 254 \text{ pounds}$$

Example 9: What force F will be required to lower a load of 6000 pounds using the screw referred to in **Example 8**? In this case, the load assists in turning the screw; hence,

$$F = 6000 \times \frac{6.2832 \times 0.150 \times 2 - 0.75}{6.2832 \times 2 + 0.150 \times 0.75} \times \frac{2}{10} = 107 \text{ pounds}$$

Coefficients of Friction for Screws and Their Efficiency.— According to experiments Professor Kingsbury made with square-threaded screws, a friction coefficient μ of 0.10 is about right for pressures less than 3000 pounds per square inch (psi) and velocities above 50 feet per minute, assuming that fair lubrication is maintained. If the pressures vary from 3000 to 10,000 psi, a coefficient of 0.15 is recommended for low velocities. The coefficient of friction varies with lubrication and the materials used for the screw and nut. For pressures of 3000 psi and by using heavy machinery oil as a lubricant, the coefficients were as follows: Mild steel screw and cast-iron nut, 0.132; mild-steel nut, 0.147; cast-brass nut, 0.127. For pressures of 10,000 psi using a mild-steel screw, the coefficients were, for a cast-iron nut, 0.136; for a mild-steel nut, 0.141; for a cast-brass nut, 0.136. For dry screws, the coefficient may be 0.3 to 0.4 or higher.

Frictional resistance is proportional to the normal pressure, and for a thread of angular form, the increase in the coefficient of friction is equivalent practically to $\mu \sec\beta$, in which β equals one-half the included thread angle; hence, for a 60-degree thread, a coefficient of 1.155μ may be used. The square form of thread has a somewhat higher efficiency than threads with sloping sides, although when the angle of the thread form is comparatively small, as in an Acme thread, there is little increase in frictional losses. Multiple-thread screws are much more efficient than single-thread screws, as the efficiency is affected by the helix angle of the thread.

The efficiency between a screw and nut increases quite rapidly for helix angles up to 10 to 15 degrees (measured from a plane perpendicular to the screw axis). The efficiency remains nearly constant for angles between about 25 and 65 degrees, and the angle of maximum efficiency is between 40 and 50 degrees. A screw will not be self-locking if the efficiency exceeds 50 percent. For example, the screw of a jack or other lifting or hoisting appliance would turn under the action of the load if the efficiency were over 50 percent. It is evident that maximum efficiency for power transmission screws often is impractical, as for example, when the smaller helix angles are required to permit moving a given load by the application of a smaller force or turning moment than would be needed for a multiple-screw thread.

In determining the efficiency of a screw and a nut, the helix angle of the thread and the coefficient of friction are the important factors. If E equals the efficiency, A equals the helix angle, measured from a plane perpendicular to the screw axis, and μ equals the coefficient of friction between the screw thread and nut, then the efficiency may be determined by the following formula, which does not take into account any additional friction losses, such as may occur between a thrust collar and its bearing surfaces:

$$E = \frac{\tan A(1 - \mu \tan A)}{\tan A + \mu}$$

This formula would be suitable for a screw having ball-bearing thrust collars. Where collar friction should be taken into account, a fair approximation may be obtained by changing the denominator of the foregoing formula to $\tan A + 2\mu$. Otherwise, the formula remains the same.

Angles and Angular Velocity Expressed in Radians.—There are three systems generally used to indicate the sizes of angles, which are ordinarily measured by the number of degrees in the arc subtended by the sides of the angle. Thus, if the arc subtended by the sides of the angle equals one-sixth of the circumference, the angle is said to be 60 degrees. Angles are also designated as multiples of a right angle. As an example, the sum of the interior angles of any polygon equals the number of sides less two, times two right angles. Thus the sum of the interior angles of an octagon equals $(8-2)\times 2\times 90=6\times 180=1080$ degrees. Hence each interior angle equals $1080\div 8=135$ degrees.

A third method of designating the size of an angle is very helpful in certain problems. This method makes use of radians. A radian is defined as a central angle, the subtended arc of which equals the radius of the arc.

By using the symbols on *Handbook* page **176**, v may represent the length of an arc as well as the velocity of a point on the periphery of a body. Then, according to the definition of a radian: $\omega = v/r$, or the angle in radians equals the length of the arc divided by the radius. Both the length of the arc and the radius must, of course, have the same unit of measurement—both must be in feet or inches or centimeters, etc. By rearranging the preceding equation:

$$v = \omega r \quad \text{and} \quad r = \frac{v}{\omega}$$

These three formulas will solve practically every problem involving radians.

The circumference of a circle equals πd or $2\pi r$, which equals $6.2832r$, which indicates that a radius is contained in a circumference 6.2832 times; hence there are 6.2832 radians in a circumference. Since a circumference represents 360 degrees, 1 radian equals $360\div 6.2832=57.2958$ degrees. Since 57.2958 degrees = 1 radian, 1 degree = 1 radian $\div$ 57.2958 = 0.01745 radian.

Example 10: 2.5 radians equal how many degrees? One radian = 57.2958 degrees; hence, 2.5 radians = $57.2958\times 2.5=143.239$ degrees.

Example 11: $22°\,31'\,12''$ = how many radians? 12 seconds = $^{12}/_{60}$ = $^{1}/_{5}$ = 0.2 minute; $31.2'\div 60=0.52$ degree. One radian = 57.3 degrees approximately. $22.52° = 22.52 + 57.3 = 0.393$ radian.

Example 12: In the figure on *Handbook* page **79**, let $l = v$ = 30 inches; and radius $r = 50$ inches; find the central angle $\omega = v/r = {}^{30}/_{50} = {}^{3}/_{5} = 0.6$ radian.

$$57.2958\times 0.6=34°22.6'$$

Example 13: $^{3\pi}/_{4}$ radians equal how many degrees? 2π radians $=360°$; π radians = 180°. $^{3\pi}/_{4} = {}^{3}/_{4}\times 180=135$ degrees.

Example 14: A 20-inch grinding wheel has a surface speed of 6000 feet per minute. What is the angular velocity?

The radius (r) = $^{10}/_{12}$ foot; the velocity (n) in feet per second = $^{6000}/_{60}$; hence,

$$\omega = \frac{6000}{60 \times {}^{10}/_{12}} = 120 \text{ radians per second}$$

Example 15: Use the table on *Handbook* page **98** to solve **Example 11**.

$$
\begin{aligned}
20° &= 0.349066 \text{ radian} \\
2° &= 0.034907 \text{ radian} \\
31' &= 0.009018 \text{ radian} \\
12'' &= 0.000058 \text{ radian} \\
\hline
22°31'12'' &= 0.393049 \text{ radian}
\end{aligned}
$$

Example 16: 7.23 radians equals how many degrees? On *Handbook* page **99**, find:

$$
\begin{aligned}
7.0 \text{ radians} &= 401°\,4'\,14'' \\
0.2 \text{ radian} &= 11°\,27'\,33'' \\
0.03 \text{ radian} &= 1°\,43'\,8'' \\
\hline
7.23 \text{ radians} &= 414°\,14'\,55''
\end{aligned}
$$

PRACTICE EXERCISES FOR SECTION 15

(See *Answers to Practice Exercises for Section 15* on page **246**)

1) In what respect does a foot-pound differ from a pound?

2) If a 100-pound weight is dropped, how much energy will it be capable of exerting after falling 10 feet?

3) Can the force of a hammer blow be expressed in pounds?

4) If a 2-pound hammer is moving 30 feet per second, what is its kinetic energy?

5) If this 2-pound hammer drives a nail into a $^{1}/_{4}$-inch board, what is the average force of the blow?

6) What relationship is there between the muzzle velocity of a projectile fired upward and the velocity with which the projectile strikes the ground?

7) What is the difference between the composition of forces and the resolution of forces?

8) If four equal forces act along lines 90 degrees apart through a given point, what is the shape of the corresponding polygon of forces?

9) Skids are to be employed for transferring boxed machinery from one floor to the floor above. If these skids are inclined at an angle of 35 degrees, what force in pounds, applied parallel to the skids, will be required to slide a boxed machine weighing 2500 pounds up the incline, assuming that the coefficient of friction is 0.20?

10) Refer to Exercise 9. If the force or pull were applied in a horizontal direction instead of in line with the skids, what increase, if any, would be required?

11) Will the boxed machine referred to in #9 above slide down the skids by gravity?

12) At what angle will the skids need to be for the boxed machine to start sliding due to gravity?

13) What name is applied to the angle that marks the dividing line between sliding and nonsliding when a body is placed on an inclined plane?

14) How is the "angle of repose" determined?

15) What figure or value is commonly used in engineering calculations for acceleration due to gravity?

16) Is the value commonly used for acceleration due to gravity strictly accurate for any locality?

17) A flywheel 3 feet in diameter has a rim speed of 1200 feet per minute, and another flywheel 6 feet in diameter has the same rim speed. Will the rim stress or the force tending to burst the larger flywheel be greater than the force in the rim of the smaller flywheel?

18) What factors of safety are commonly used in designing flywheels?

19) Does the stress in the rim of a flywheel increase in proportion to the rim velocity?

20) What is generally considered the maximum safe speed for the rim of a solid or one-piece cast-iron flywheel?

21) Why is a well-constructed wood flywheel better adapted to higher speeds than one made of cast iron?

22) What is the meaning of the term "critical speed" as applied to a rotating body?

23) How is angular velocity generally expressed?

24) What is a radian, and how is its angle indicated?

25) How many degrees are there in 2.82 radians?

26) How many degrees are in the following radians: $\pi/3$ and $2\pi/5$.

27) Reduce to radians: 63°; 45°32′; 6°37′46″; 22°22′22″.

28) Find the angular velocity in radians per second of the following: 157 rpm; 275 rpm; 324 rpm.

29) Why do the values in the *l* column starting on *Handbook* page **79** equal those in the radian column on page **98**?

30) If the length of the arc of a sector is 4⅞ inches, and the radius is 6⅞ inches, find the central angle.

31) A 12-inch grinding wheel has a surface speed of a mile a minute. Find its angular velocity and its revolutions per minute.

32) The radius of a circle is 1½ inches, and the central angle is 60 degrees. Find the length of the arc.

33) If an angle of 34°12′ subtends an arc of 16.25 inches, find the radius of the arc.

SECTION 16

STRENGTH OF MATERIALS

Machinery's Handbook pages **200–221**

The *Strength of Materials* section of *Machinery's Handbook* contains fundamental formulas and data for use in proportioning parts that are common to almost every type of machine or mechanical structure. In designing machine parts, factors other than strength often are of vital importance. For example, some parts are made much larger than required for strength alone to resist extreme vibrations, deflection, or wear; consequently, many machine parts cannot be designed merely by mathematical or strength calculations, and their proportions should, if possible, be based upon experience or upon similar designs that have proved successful. It is evident that no engineering handbook can take into account the endless variety of requirements relating to all types of mechanical apparatus, and it is necessary for the designer to determine these local requirements for each, but, even when the strength factor is secondary due to some other requirement, the strength, especially of the more important parts, should be calculated, in many instances, merely to prove that it will be sufficient.

In designing for strength, the part is so proportioned that the maximum working stress likely to be encountered will not exceed the strength of the material by a suitable margin. The design is accomplished by the use of a factor of safety. The relationship between the working stress s_w, the strength of the material, S_m, and the factor of safety, f_s is given by **Equation (1)** on page **205** of the *Handbook*:

$$s_w = \frac{S_m}{f_s} \qquad \text{(a)}$$

The value selected for the strength of the material, S_m depends on the type of material, whether failure is expected to occur because of tensile, compressive, or shear stress, and on whether the stresses are constant, fluctuating, or are abruptly applied as with shock loading. In general, the value of S_m is based on yield strength for ductile materials, ultimate strength for brittle materials, and fatigue strength for parts subject to cyclic stresses. Moreover, the value for S_m must be for the temperature at which the part operates. Values of S_m for common materials at 68° F can be obtained from the tables in *Machinery's Handbook* from pages **427** and **508**. Factors from the table given on *Handbook* page **390**, *Influence of Temperature on the Strength of Metals*, can be used to convert strength values at 68° F to values applicable at elevated temperatures. For heat-treated carbon and alloy-steel parts, see data starting on *Handbook* page **421**.

The factor of safety depends on the relative importance of reliability, weight, and cost. General recommendations are given in the *Handbook* on page **205**.

Working stress is dependent on the shape of the part, hence on a stress concentration factor, and on a nominal stress associated with the way in which the part is loaded. Equations and data for calculating nominal stresses, stress concentration factors, and working stresses are given starting on *Handbook* page **206**.

Example 1: Determine the allowable working stress for a part that is to be made from SAE 1112 free-cutting steel; the part is loaded in such a way that failure is expected to occur in tension when the yield strength has been exceeded. A factor of safety of 3 is to be used.

From the table, *Strength Data for Iron and Steel*, on page **427** of the *Handbook*, a value of 30,000 pounds per square inch (psi) is selected for the strength of the material, S_m. Working stress s_w is calculated from **Equation (a)** as follows:

$$s_w = \frac{30{,}000}{3} = 10{,}000 \text{ psi}$$

Finding Diameter of a Bar to Resist Safely Under a Given Load.—Assume that a direct tension load, F, is applied to a bar such that the force acts along the longitudinal axis of the bar. From *Handbook* page **210**, the following equation is given for calculating the nominal stress:

$$\sigma = \frac{F}{A} \quad \text{(b)}$$

where A is the cross-sectional area of the bar. **Equation (2)** on *Handbook* page **206** related the nominal stress to the stress concentration factor, K, and working stress, s_w:

$$s_w = K\sigma \quad \text{(c)}$$

Combining **Equation (a)**, **(b)**, and **(c)** results in the following:

$$\frac{S_m}{Kf_s} = \frac{F}{A} \quad \text{(d)}$$

Example 2: A structural steel bar supports in tension a load of 40,000 pounds. The load is gradually applied and then, after having reached its maximum value, is gradually removed. Find the diameter of round bar required.

According to the table on *Handbook* page **427**, the yield strength of structural steel is 33,000 psi. Suppose that a factor of safety of 3 and a stress concentration factor of 1.1 are used. Then, inserting known values in **Equation (d)**:

$$\frac{33{,}000}{1.1 \times 3} = \frac{40{,}000}{A}; \; A = \frac{40{,}000 \times 3.3}{33{,}000}; \; A = 4 \text{ square inches}$$

Hence, the cross-section of the bar must be about 4 square inches. As the bar is circular in section, the diameter must then be about $2\frac{1}{4}$ inches.

Diameter of a Bar to Resist Compression.—If a short bar is subjected to compression in such a way that the line of application of the load coincides with the longitudinal axis of the bar, the formula for nominal stress is the same as for direct tension loading. **Equation (b)** and hence **Equation (d)** also may be applied to direct compression loading.

Example 3: A short structural-steel bar supports in compression a load of 40,000 pounds. (See **Fig. 1**.) The load is steady. Find the diameter of the bar required.

From page **427** in the *Handbook*, the yield strength of structural steel is 33,000 psi. If a stress concentration factor of 1.1 and a factor of safety of 2.5 are used, then, substituting values into **Equation (d)**:

$$\frac{33{,}000}{1.1 \times 2.5} = \frac{40{,}000}{A}; A = 3.33 \text{ square inches}$$

The diameter of a bar, the cross-section of which is 3.33 square inches, is about $2\frac{1}{16}$ inches.

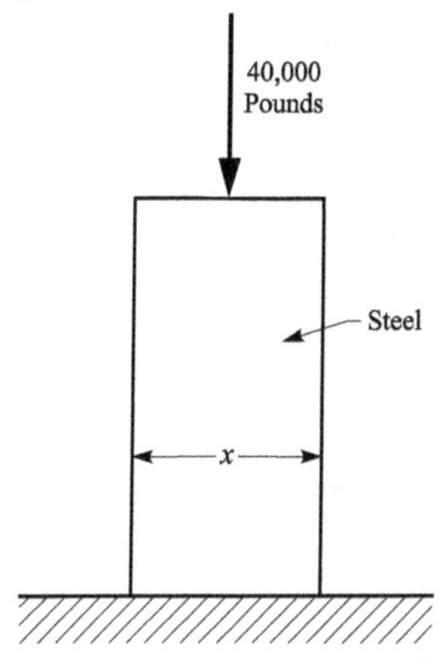

Fig. 1. Calculating Diameter *x* to Support a Given Load Safely

According to a general rule, the simple formulas that apply to compression should be used only if the length of the member being compressed is not greater than 6 times the least cross-sectional dimension. For example, these formulas should be applied to round bars only when the length of the bar is less than 6 times the diameter. If the bar is rectangular, the formulas should be applied only to bars having a length less than 6 times the shortest side of the rectangle. When bars are longer than this, a compressive stress causes a sidewise bending action, and an even distribution of the compression stresses over the total area of the cross-section should no longer be depended upon. Special formulas for long bars or columns will be found on *Handbook* page **277**; see also text beginning on page **275**, *Strength of Columns or Struts*.

Diameter of a Pin to Resist Shearing Stress.—The pin *E* shown in the illustration, **Fig. 2**, is subjected to shear. Parts *G* and *B* are held together by the pin and tend to shear it off at C and D. The areas resisting the shearing action are equal to the pin at these points.

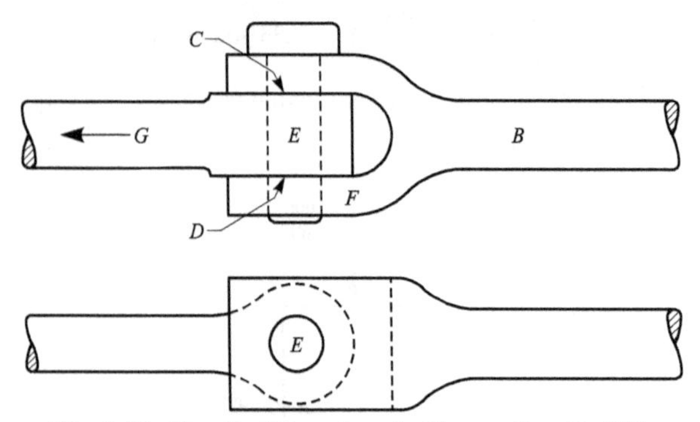

Fig. 2. Finding the Diameter of a Connecting-Rod Pin to Resist a Known Load G

From **Table 2** on pages **210–211** of the *Handbook*, the equation for direct shear is:

$$\tau = \frac{F}{A} \qquad \text{(e)}$$

τ is a simple stress related to the working stress, s_w, by **Equation (3)** on *Handbook* page **206**:

$$s_w = K\tau \qquad \text{(f)}$$

where K is a stress concentration factor. Combining **Equation (a)**, **(e)**, and **(f)** gives **Equation (d)** on page **146**, where S_m is, of course, the shearing strength of the material.

If a pin is subjected to shear as in **Fig. 2**, so that two surfaces, as at C and D, must fail by shearing before breakage occurs, the areas of both surfaces must be taken into consideration when calculating the strength. The pin is then said to be in *double shear*. If the lower part F of connecting rod B were removed, so that member G were connected with B by a pin subjected to shear at C only, the pin would be said to be in *single shear*.

Example 4: Assume that in **Fig. 2** the load at G pulling on the connecting rod is 20,000 pounds. The material of the pin is SAE

1025 steel. The load is applied in such a manner that shocks are liable to occur. Find the required dimensions for the pin.

Since the pins are subjected to shock loading, the nominal stress resulting from the application of the 20,000-pound load must be assumed to be twice as great (see *Handbook* starting on page **272**) as it would be if the load were gradually applied or steady. From *Handbook* page **427**, the ultimate strength in shear for SAE 1025 steel is 75 percent of 60,000 or 45,000 psi. A factor of safety of 3 and a stress concentration factor of 1.8 are to be used. By substituting values into **Equation (d)**:

$$\frac{45{,}000}{1.8 \times 3} = \frac{2 \times 20{,}000}{A}; \quad A = \frac{10.8 \times 20{,}000}{45{,}000}$$

$$= 4.8 \text{ square inch}$$

As the pin is in double shear, that is, as there are two surfaces *C* and *D* over which the shearing stress is distributed, each surface must have an area of one-half the total shearing area *A*. Then, the cross-sectional area of the pin will be 2.4 square inches, and the diameter of the pin, to give a cross-sectional area of 2.4 square inches, must be $1\frac{3}{4}$ inches.

Beams and Stresses to Which They Are Subjected.—Parts of machines and structures subjected to bending are known mechanically as *beams*. Hence, in this sense, a lever fixed at one end and subjected to a force at its other end, a rod supported at both ends and subjected to a load at its center, or the overhanging arm of a jib crane would all be known as beams.

The stresses in a beam are principally tension and compression stresses. If a beam is supported at the ends, and a load rests upon the upper side, the lower fibers will be stretched by the bending action and will be subjected to a tensile stress, while the upper fibers will be compressed and be subjected to a compressive stress. There will be a slight lengthening of the fibers in the lower part of the beam, while those on the upper side will be somewhat shorter, depending upon the amount of deflection. If we assume that the beam is either round or square in cross-section, there will be a layer or surface through its center line, which will be neither in compression nor in tension.

This surface is known as the neutral surface. The stresses of the individual layers or fibers of the beam will be proportional to their distances from the neutral surface, the stresses being greater the farther away from the neutral surface the fiber is located. Hence, there is no stress on the fibers in the neutral surface, but there is a maximum tension on the fibers at the extreme lower side and a maximum compression on the fibers at the extreme upper side of the beam. In calculating the strength of beams, it is, therefore, only necessary to determine that the fibers of the beam that are at the greatest distance from the neutral surface are not stressed beyond the safe working stress of the material. If this condition exists, all the other parts of the section of the beam are not stressed beyond the safe working stress of the material.

In addition to the tension and compression stresses, a loaded beam is also subjected to a stress that tends to shear it. This shearing stress depends upon the magnitude and kind of load. In most instances, the shearing action can be ignored for metal beams, especially if the beams are long and the loads far from the supports. If the beams are very short and the load quite close to a support, then the shearing stress may become equal to or greater than the tension or compression stresses in the beam, and the beam should then be calculated for shear.

Beam Formulas.—The bending action of a load upon a beam is called the *bending moment*. For example, in **Fig. 3** the load P acting downward on the free end of the cantilever beam has a moment or bending action about the support at A equal to the load multiplied by its distance from the support. The bending moment is commonly expressed in inch-pounds, the load being expressed in pounds and the lever arm or distance from the support in inches. The length of the lever arm should always be measured in a direction at right angles to the direction of the load. Thus, in the second example in **Fig. 3**, the bending moment is not $P \times a$, but is $P \times l$, because l is measured in a direction at right angles to the direction of the load P.

The property of a beam to resist the bending action or the bending moment is called the *moment of resistance* of the beam. It is evident that the bending moment must be equal to the moment of resistance. The moment of resistance, in turn, is equal to the stress in the fiber farthest away from the neutral plane multiplied by the section modulus. The *section modulus* is a factor that depends upon the shape and size of the cross-section of a beam and is given for different cross-sections in all engineering handbooks. (See table, *Moments of Inertia, Section Moduli, and Radii of Gyration* starting on *Handbook* page **235**.) The section modulus, in turn, equals the moment of inertia of the cross-section, divided by the distance from the neutral surface to the most extreme fiber. The moment of inertia formulas for various cross-sections also will be found in the table just mentioned.

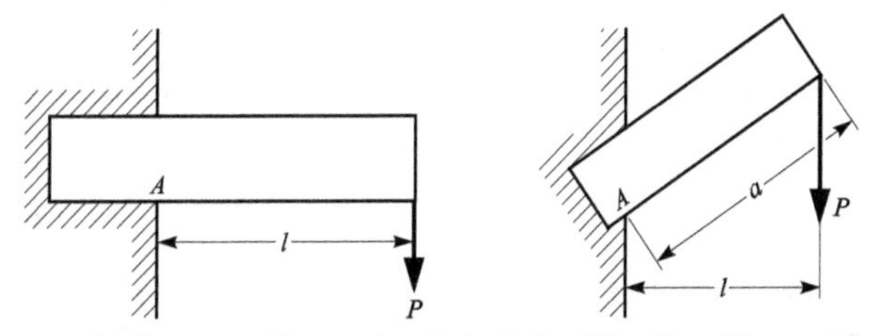

Fig. 3. Diagrams Illustrating Principle of Bending Moments

The following formula on *Handbook* page **211** may be given as the fundamental formula for bending of beams:

$$\sigma = \pm\frac{M}{Z} = \pm\frac{My}{I} \tag{g}$$

The moment of inertia I is a property of the cross-section that determines its relative strength. In calculations of strength of materials, a handbook is necessary because of the tabulated formulas and data relating to section moduli and moments of inertia, areas of cross-sections, etc., to be found therein.

There are many different ways in which a beam can be supported and loaded, and the bending moment caused by a given load varies greatly according to whether the beam is supported at one end only or at both ends, also whether it is freely supported at the ends or is held firmly. The load may be equally distributed over the full length of the beam or may be applied at one point either in the center or near to one or the other of the supports. The point where stress is maximum is generally called the critical point. The stress at the critical point equals bending moment divided by section modulus.

Formulas for determining the stresses at the critical points will be found in the table of beam formulas, starting on *Handbook* page **251**.

Example 5: A rectangular steel bar 2 inches thick and firmly built into a wall, as shown in **Fig. 4**, is to support 3000 pounds at its outer end 36 inches from the wall. What would be the necessary depth h of the beam to support this weight safely?

The bending moment equals the load times the distance from the point of support, or $3000 \times 36 = 108{,}000$ inch-pounds.

By combining **Equation (a)**, **(c)**, and **(g)**, the following equation is obtained:

$$\frac{S_m}{Kf_s} = \frac{M}{Z} \qquad \text{(h)}$$

If the beam is made from structural steel, the value for S_m, based on yield strength, from page **427** in the *Handbook*, is 33,000 psi. By using a stress concentration factor of 1.1 and a factor of safety of 2.5, values may be inserted into the above equation:

$$\frac{33{,}000}{1.1 \times 2.5} = \frac{108{,}000}{Z}; \quad Z = \frac{2.75 \times 108{,}000}{33{,}000}; \quad Z = 9 \text{ inches}^3$$

The section modulus for a rectangle equals $bd^2/6$, in which b is the length of the shorter side and d of the longer side of the rectangle (see *Handbook* page **236**), hence, $Z = bd^2/6$.

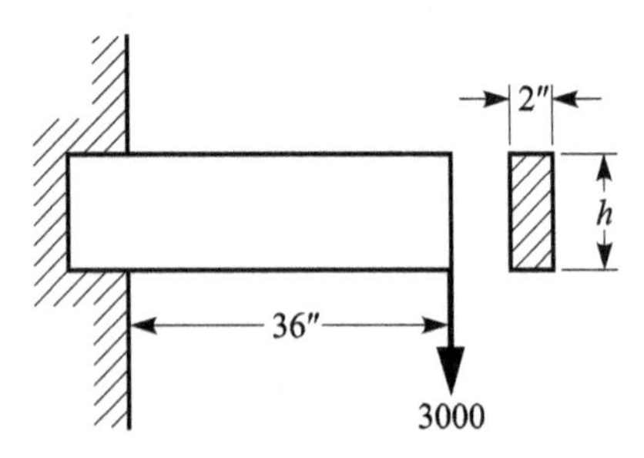

Fig. 4. Determining the Depth *h* of a Beam to Support a Known Weight

But $Z = 9$ and $b = 2$. Inserting these values into the formula, we have:

$$9 = \frac{2d^2}{6}$$

from which $d^2 = 27$, and $d = 5.2$ inches. This value d corresponds to dimension h in **Fig. 4**. Hence, the required depth of the beam to support a load of 3000 pounds at the outer end with a factor of safety of 3 would be 5.2 inches.

In calculating beams having either rectangular or circular cross-sections, the formulas on *Handbook* pages **262–263** are convenient to use. A beam loaded as shown by **Fig. 4** is similar to the first diagram on *Handbook* page **262**. If the formula on this page in the *Handbook* for determining height h is applied to **Example 5**, **Fig. 4**, then,

$$h = \sqrt{\frac{6lW}{bf}} \qquad \sqrt{\frac{6 \times 36 \times 3000}{2 \times 12{,}000}} = 5.2 \text{ inches}$$

In the above calculation the stress value f is equivalent to S_m/Kf_s.

Example 6: A steel I-beam is to be used as a crane trolley track. This I-beam is to be supported at the ends, and the unsupported span is 20 feet long. The maximum load is 6000 pounds, and the nominal stress is not to exceed 10,000 pounds per square inch. Determine the size of the standard I-beam; also determine the maximum deflection when the load is at the center of the beam.

The foregoing conditions are represented by Case 2, *Handbook* page **251**. A formula for the stress at the critical point is $Wl/4Z$. As explained on *Handbook* page **250**, all dimensions are in inches, and the minus sign preceding a formula merely denotes compression of the upper fibers and tension in the lower fibers.

By inserting the known values in the formula:

$$10{,}000 = \frac{6000 \times 240}{4Z}; \text{ hence}$$

$$Z = \frac{6000 \times 240}{10{,}000 \times 4} = 36$$

The table of standard I-beams on *Handbook* page **2702** shows that a 12-inch I-beam, which weighs 31.8 pounds per foot, has a section modulus of 36.4.

The formula for maximum deflection (see *Handbook* starting on page **251**, Case 2) is $Wl^3/48EI$. According to the table on *Handbook* page **427**, the modulus of elasticity (E) of structural steel is 29,000,000.

As Z = moment of inertia I ÷ distance from neutral axis to extreme fiber (see *Handbook* page **250**), then for a 12-inch I-beam $I = 6Z = 216$; hence,

$$\text{Maximum deflection} = \frac{6000 \times (240)^3}{48 \times 29{,}000{,}000 \times 216} = 0.27 \text{ inch}$$

Example 7: All conditions are the same as in **Example 6**, except that the maximum deflection at the "critical point," or center of the I-beam, must not exceed $\frac{1}{8}$ inch. What size I-beam is required?

To meet the requirement regarding deflection,

$$\frac{1}{8} = \frac{Wl^3}{48EI}; \quad \text{therefore,}$$

$$I = \frac{8Wl^3}{48E} = \frac{8 \times 6000 \times (240)^3}{48 \times 29{,}000{,}000} = 476$$

If x = distance from neutral axis to most remote fiber ($\frac{1}{2}$ beam depth in this case), then $Z = I/x$, and the table on *Handbook* page **2702** shows that a 15-inch, 50-pound I-beam should be used because it has a section modulus of 64.8 and 476/7.5 = 63.5 nearly.

If 476 were divided by 6 ($\frac{1}{2}$ depth of a 12-inch I-beam), the result would be much higher than the section modulus of any standard 12-inch I-beam (476 ÷ 6 = 79.3); moreover, 576 ÷ 9 = 53, which shows that an 18-inch I-beam is larger than is necessary because the lightest beam of this size has a section modulus of 81.9.

Example 8: If the speed of a motor is 1200 revolutions per minute and if its driving pinion has a pitch diameter of 3 inches, determine the torsional moment to which the pinion shaft is subjected, assuming that 10 horsepower is being transmitted.

If W = tangential load in pounds, H = the number of horsepower, and V = pitch-line velocity in feet per minute,

$$W = \frac{33,000 \times H}{V}$$
$$= \frac{33,000 \times 10}{943} = 350 \text{ pounds}$$

The torsional moment = $W \times$ pitch radius of pinion = 350×1.5 = 525 pound-inches (or inch-pounds).

Example 9: If the pinion referred to in **Example 8** drives a gear having a pitch diameter of 12 inches, to what torsional or turning moment is the gear shaft subjected?

The torque or torsional moment in any case = pitch radius of gear × tangential load. The latter is the same for both gear and pinion; hence, torsional moment of gear = $350 \times 6 = 2100$ inch-pounds.

The torsional moment or the turning effect of a force that tends to produce rotation depends upon (1) the magnitude of the force acting, and (2) the distance of the force from the axis of rotation, measuring along a line at right angles to the line of action of the force.

Strength in Plastic and Polymer Composite Parts.—Various plastic materials are broadly used and provide desirable properties. Commonly used plastics, such as polyurethane and silicone, tend to be *viscoelastic*, displaying properties between those of elastic solids and viscous fluids, exhibiting a nonlinear stress-strain relationship. When these materials are stressed by sustained loads, they often deform or *creep*; composition, exposure time, and temperature affect the rate at which this can occur. After such deformation, the object will not return to its original shape and dimensions, causing a reduction in material strength. For more on mechanical, thermal, and electrical properties of plastics, see *Handbook* pages **561–580**; manufacturing for plastics is discussed on pages **580–591**; plastics parts design is discussed on pages **591–618**.

Polymer Composites: For applications requiring higher strength-to-weight ratios, good corrosion resistance, and reduced vibrational effects, products may be engineered with a matrix containing oriented composite plies or other polymer fibers for added reinforcement. The majority of the strength and stiffness of most composites is provided by the fiber reinforcements, while the matrix is a continuous material that holds the fiber reinforcements together. Typical loadings of fiber reinforcements are on the order of 50 percent by component volume and 75 percent by component mass, with the balance going to the matrix material.

Formulas for calculating density, maximum modulus, and maximum strength of polymer composites can be found on *Handbook* page **586**.

A composite's material properties can be estimated by considering the volume fraction (f) of the fiber reinforcements, as well as the orientation of the fiber reinforcements relative to the load direction. The density (ρ_C), maximum modulus (E_C), and maximum strength (σ_C) of the composite in the direction of the reinforcements are estimated using a rule of mixtures, where the properties for the reinforcements and matrix are indicated by the subscripts R and M, respectively:

$$\begin{aligned} \rho_C &= f\rho_R + (1-f)\,\rho_M \\ E_C &= fE_R + (1-f)\,E_M \\ \sigma_C &= f\sigma_R + (1-f)\,\sigma_M \end{aligned}$$

Example 10: Use these formulas to determine the density, maximum modulus, and maximum strength of a composite with 55 percent volume fraction of carbon fibers embedded within an epoxy thermoset matrix:

Carbon properties: density = 1.81g/cm^3, tensile modulus = 242 GPa, tensile strength = 4.1 GPa

Epoxy properties: density = 1.35g/cm^3, tensile modulus = 35 GPa, tensile strength = 0.56 GPa

$$\rho_C = 0.55 \times \frac{1.81g}{cm^3} + (1 - 0.55) \times \frac{1.35g}{cm^3} = \frac{1.60g}{cm^3}$$

$$E_C = 0.55 \times 242\ GPa + (1 - 0.55) \times 35\ GPa = 148.85\ GPa$$

$$\sigma_C = 0.55 \times 4.1\ GPa + (1 - 0.55) \times 0.56\ GPa = 2.51\ GPa$$

Sustainability Considerations.—Added to the challenges of selecting materials with the appropriate mechanical and thermal strength and durability, fine-tuning design parameters, and optimizing manufacturing and machining methods, is the ever more important emphasis on sustainable materials and processes. For an overview of some common materials and calculations related to energy consumption, production cost, carbon footprint, and more, see pages **361–366** of the *Handbook*.

PRACTICE EXERCISES FOR SECTION 16

(See *Answers to Practice Exercises for Section 16* on page **248**)

1) What is a "factor of safety," and why are different factors used in machine design?

2) If the ultimate strength of a steel rod is 60,000 pounds per square inch, and the factor of safety is 5, what is the equivalent working stress?

3) If a steel bar must withstand a maximum pull of 9000 pounds and if the maximum nominal stress must not exceed 12,000 pounds per square inch, what diameter bar is required?

4) Is a steel rod stronger when at ordinary room temperature or when heated to 500° F?

5) What is the meaning of the term "elastic limit"?

6) Approximately what percentages of copper and zinc in brass result in the greatest tensile strength?

7) If four 10-foot-long pipes are to be used to support a water tank installation weighing 100,000 pounds, what diameter standard-weight pipe is required?

8) Under the elastic limit of common metals, how is stress correlated to strain?

9) When plastics are exhibiting high loads, what common phenomena occurs? What happens when the temperature also increases with this high loading?

10) What are the two major components of polymer composites? What component provides the tensile strength?

SECTION 17

DESIGN OF SHAFTS AND KEYS FOR POWER TRANSMISSION

Machinery's Handbook pages **289–297** and pages **2545–2568**

This section is a review of the general procedure in designing shafts to resist both torsional and combined torsional and bending stresses. The diameter of a shaft through which power is transmitted depends, for a given shaft material, upon the amount and kind of stress or stresses to which the shaft is subjected. To illustrate the general procedure, we shall assume first that the shaft is subjected only to a uniform torsional or twisting stress and that there is no additional bending stress that needs to be considered in determining the diameter.

Example 1: A lineshaft carrying pulleys located close to the bearings is to transmit 50 horsepower at 1200 revolutions per minute. If the load is applied gradually and is steady, what diameter steel shaft is required, assuming that the pulleys are fastened to the shaft by means of keys and that the bending stresses caused by the pull of the belts are negligible?

According to the former American Standard Association's Code for the Design of Transmission Shafting, the diameter of shaft required to meet the stated conditions can be determined by using **Formula (16b)**, *Handbook* page **294**.

$$D = B \times \sqrt[3]{\frac{321,000K_tP}{S_sN}}$$

In this formula, D = required shaft diameter in inches; B = a factor, which for solid shafts is taken as 1; K_t = combined shock and fatigue factor; P = maximum horsepower transmitted by shaft; S_s = maximum allowable torsional shearing stress in pounds per square inch; and N = shaft speed in revolutions per minute.

From **Table 1** on *Handbook* page **295**, K_t = 1.0 for gradually applied and steady loads, and from **Table 2** (on the same page) the recommended maximum allowable working stress for "Commercial Steel" shafting with keyways subjected to pure torsion loads is 6000 pounds per square inch. By substituting in the formula,

$$D = 1 \times \sqrt[3]{\frac{321{,}000 \times 1.0 \times 50}{6000 \times 1200}} = 1.306 \text{ inches}$$

The nearest standard-size transmission shafting from the table on *Handbook* page **293** is $1\frac{7}{16}$ inches.

Example 2: If, in **Example 1**, the shaft diameter had been determined by using **Formula (5b)**, *Handbook* page **289**, what would the result have been and why?

$$D = \sqrt[3]{\frac{53.5P}{N}} = \sqrt[3]{\frac{53.5 \times 50}{1200}} = 1.306 \text{ inches}$$

This formula gives the same shaft diameter as was previously determined because it is a simplified form of the first formula used and contains the same values of K_t and S_s, but combined as the single constant 53.5. For lineshafts carrying pulleys under conditions ordinarily encountered, this simplified formula is usually quite satisfactory; but, where conditions of shock loading are known to exist, it is safer to use **Formula (16b)**, *Handbook* page **294**, which takes such conditions into account.

Shafts Subjected to Combined Stresses.—The preceding formulas are based on the assumption that the shaft is subjected to torsional stresses only. However, many shafts must withstand stresses that result from combinations of torsion, bending, and shock loading. In such conditions it is necessary to use formulas that take such stresses into account.

Example 3: Suppose that, after the lineshaft in **Example 1** was installed, it became necessary to relocate a machine that was being driven by one of the pulleys on the shaft. Because of the new machine location, it was necessary to move the pulley on the lineshaft farther away from the nearest bearing, and, as a result, a bending moment of 2000 inch-pounds was introduced. Is the

$1\frac{7}{16}$-inch diameter shaft sufficient to take this additional stress, or will it be necessary to relocate the bearing to provide better support?

Since there are now both bending and torsional loads acting on the shaft, **Formula (18b)**, *Handbook* page **294** should be used to compute the required shaft diameter. This diameter is then compared with the $1\frac{7}{16}$-inch diameter previously determined.

$$D = B \times \sqrt[3]{\frac{5.1}{p_t}\sqrt{(K_m M)^2 + \left(\frac{63{,}000 K_t P}{N}\right)^2}}$$

In this formula B, K_t, P, and N are quantities previously defined and p_t = maximum allowable shearing stress under combined loading conditions in pounds per square inch; K_m = combined shock and fatigue factor; and M = maximum bending moment in inch-pounds.

From **Table 1** on *Handbook* page **295**, K_m = 1.5 for gradually applied and steady loads and from **Table 2** (on the same page), p_t = 6000 pounds per square inch. By substituting in the formula,

$$D = 1 \times \sqrt[3]{\frac{5.1}{6000}\sqrt{(1.5 \times 2000)^2 + \left(\frac{63{,}000 \times 1 \times 50}{1200}\right)^2}}$$

$$= \sqrt[3]{\frac{5.1}{6000}\sqrt{9000000 + 6{,}890{,}625}} = \sqrt[3]{\frac{5.1}{6000} \times 3986}$$

$$= \sqrt[3]{3.388} = 1.502 \text{ inches or about } 1\tfrac{1}{2} \text{ inches}$$

This diameter is larger than the $1\frac{7}{16}$-inch diameter used for the shaft in **Example 1**, so it will be necessary to relocate the bearing closer to the pulley, thus reducing the bending moment. The $1\frac{7}{16}$-inch diameter shaft will then be able to operate within the allowable working stress for which it was originally designed.

Design of Shafts to Resist Torsional Deflection.—Shafts must often be proportioned not only to provide the strength required to transmit a given torque, but also to prevent torsional deflection (twisting) through a greater angle than has been found satisfactory for a given type of service. This requirement is particularly true for machine shafts and machine tool spindles.

For ordinary service, it is customary that the angle of twist of machine shafts be limited to $\frac{1}{10}$ degree per foot of shaft length, and, for machine shafts subject to load reversals, $\frac{1}{20}$ degree per foot of shaft length. As explained in the *Handbook*, the usual design procedure for shafting that is to have a specified maximum angular deflection is to compute the diameter of shaft required based on both deflection and strength considerations and then to choose the larger of the two diameters thus determined.

Example 4: A 6-foot-long feed shaft is to transmit a torque of 200 inch-pounds. If there are no bending stresses, and the shaft is to be limited to a torsional deflection of $\frac{1}{20}$ degree per foot of length, what diameter shaft should be used? The shaft is to be made of cold drawn steel and is to be designed for a maximum working stress of 6000 pounds per square inch in torsion.

The diameter of shaft required for a maximum angular deflection α is given by **Formula (13)**, *Handbook* page **291**.

$$D = 4.9 \sqrt[4]{\frac{Tl}{G\alpha}}$$

In this formula T = applied torque in inch-pounds; l = length of shaft in inches; G = torsional modulus of elasticity, which, for steel, is 11,500,000 pounds per square inch; and α = angular deflection of shaft in degrees.

In the problem at hand, $T = 200$ inch-pounds; $l = 6 \times 12 = 72$ inches; and $\alpha = 6 \times 1/20 = 0.3$ degree.

$$D = 4.9 \sqrt[4]{\frac{200 \times 72}{11{,}500{,}000 \times 0.3}} = 4.9 \sqrt[4]{0.0041739}$$

$$= 4.9 \times 0.254 = 1.24 \text{ inches}$$

The diameter of the shaft based on strength considerations is obtained by using **Formula (3a)**, *Handbook* page **289**.

$$D = \sqrt[3]{\frac{5.1T}{S_s}} = \sqrt[3]{\frac{5.1 \times 200}{6000}} = \sqrt[3]{0.17} = 0.55 \text{ inch}$$

From the above calculations, the diameter based on torsional deflection considerations is the larger of the two values obtained, so the nearest standard diameter, $1\frac{1}{4}$ inches, should be used.

Selection of Key Size Based on Shaft Size.—Keys are generally proportioned in relation to shaft diameter instead of in relation to torsional load to be transmitted because of practical reasons, such as standardization of keys and shafts. Standard sizes are listed in the table *Key Size versus Shaft Diameter ANSI/ASME B17.1-1967 (R2013)* on *Handbook* page **2557**. Dimensions of both square and rectangular keys are given, but for shaft diameters up to and including $6\frac{1}{2}$ inches, square keys are preferred. For larger shafts, rectangular keys are commonly used.

Two rules that base key length on shaft size are: (1) $L = 1.5D$ and (2) $L = 0.3D^2 \div T$, where L = length of key, D = diameter of shaft, and T = key thickness.

If the keyset is to have fillets, and the key is to be chamfered, suggested dimensions for these modifications are given on *Handbook* page **2562**. If a set screw is to be used over the key, suggested sizes are given in the table on *Handbook* page **2562**.

Example 5: If the maximum torque output of a 2-inch diameter shaft is to be transmitted to a keyed pulley, what should be the proportions of the key?

According to the table on *Handbook* page **2557**, a $\frac{1}{2}$-inch square key would be preferred. If a rectangular key were selected, its dimensions would be $\frac{1}{2}$ inch by $\frac{3}{8}$ inch. According to rule 1 above, its length would be 3 inches.

The key and keyseat may be proportioned so as to provide a clearance or an interference fit. The table on *Handbook* page **2561** gives tolerances for widths and depths of keys and keyseats to provide Class 1 (clearance) and Class 2 (interference) fits. An additional Class 3 (interference) fit, which has not been standardized, is mentioned on *Handbook* page **2557** together with suggested tolerances.

Keys Proportioned According to Transmitted Torque.—As previously stated, if key sizes are based on shaft diameter, the dimensions of the key sometimes will be excessive, usually when a gear or pulley transmits only a portion of the total torque capacity of the shaft to which it is keyed. If excessively large keys are to be avoided, it may be advantageous to base the determination on the torque to be transmitted rather than on the shaft diameter and to use the dimensions thus determined as a guide in selecting a standard-size key.

A key proportioned to transmit a specified torque may fail in service either by shearing or by crushing, depending on the proportions of the key and the manner in which it is fitted to the shaft and hub. The best proportions for a key are those that make it equally resistant to failure by shearing and by crushing. The safe torque in inch-pounds that a key will transmit, based on the allowable shearing stress of the key material, may be found from the formula:

$$T_s = L \times W \times \frac{D}{2} \times S_s \quad (1)$$

The safe torque based on the allowable compressive stress of the key material is found from the formula:

$$T_c = L \times \frac{H}{2} \times \frac{D}{2} \times S_c \quad (2)$$

(For Woodruff keys the amount that the key projects above the shaft is substituted for $H/2$.)

In these formulas, T_s = safe torque in shear; T_c = safe torque in compression; S_s = allowable shearing stress; S_c = allowable compressive stress; L = key length in inches; W = key width in inches; H = key thickness in inches; and D = shaft diameter in inches.

To satisfy the condition that the key be equally resistant to shearing and crushing, T_s should equal T_c. Thus, by equating **Formulas (1)** and **(2)**, it is found that the width of the keyway in terms of the height of the keyway is:

$$W = \frac{HS_c}{2S_s} \quad (3)$$

For the type of steel commonly used in making keys, the allowable compressive stress Sc may be taken as twice the allowable shearing stress Ss of the material if the key is properly fitted on all four sides. By substituting $Sc = 2Ss$ in **Formula (3)** it will be found that $W = H$, so that for equal strength in compression and shear a square key should be used.

If a rectangular key is used, and the thickness H is less than the width W, then the key will be weaker in compression than in shear so that it is sufficient to check the torque capacity of the key using **Formula (2)**.

Example 6: A 3-inch shaft is to deliver 100 horsepower at 200 revolutions per minute through a gear keyed to the shaft. If the hub of the gear is 4 inches long, what size key, equally strong in shear and compression, should be used? The allowable compressive stress in the shaft is not to exceed 16,000 pounds per square inch and the key material has an allowable compressive stress of 20,000 pounds per square inch and an allowable shearing stress of 15,000 pounds per square inch.

The first step is to decide on the length of the key. Since the hub of the gear is 4 inches long, a key of the same length may be used. The next step is to determine the torque that the key will have to transmit. By using **Formula (2)**, *Handbook* page **289**,

$$T = \frac{63{,}000P}{N} = \frac{63{,}000 \times 100}{200} = 31{,}500 \text{ inch-pounds}$$

To determine the width of the key, based on the allowable shearing stress of the key material, **Equation (1)** above is used.

$$T_s = L \times W \times \frac{D}{2} \times S_s$$

$$31{,}500 = 4 \times W \times \frac{D}{2} \times 15{,}000$$

or

$$W = \frac{31{,}500 \times 2}{15{,}000 \times 4 \times 3} = 0.350, \text{ say, } \tfrac{3}{8} \text{ inch}$$

In using **Equation (2)** to determine the thickness of the key, however, it should be noted that, if the shaft material has a different allowable compressive stress than the key material, then the lower of the two values should be used. The shaft material then has the lower allowable compressive stress, and the keyway in the shaft would fail by crushing before the key would fail. Therefore,

$$T_c = L \times \frac{H}{2} \times \frac{D}{2} \times S_c$$

$$31{,}500 = 4 \times \frac{H}{2} \times \frac{3}{2} \times 16{,}000$$

$$H = \frac{31{,}500 \times 2 \times 2}{16{,}000 \times 4 \times 3} = 0.656 = \frac{21}{32} \text{ inch}$$

Therefore, the dimensions of the key for equal resistance to failure by shearing and crushing are $^{3}/_{8}$ inch wide, $^{21}/_{32}$ inch thick, and 4 inches long. If, for some reason, it is desirable to use a key shorter than 4 inches, say, 2 inches, then it will be necessary to increase both the width and thickness by a factor of 4 ÷ 2 if equal resistance to shearing and crushing is to be maintained. Thus the width would be $^{3}/_{8} \times {}^{4}/_{2} = {}^{3}/_{4}$ inch, and the thickness would be $^{21}/_{32} \times {}^{4}/_{2} = 1^{5}/_{16}$ inch for a 2-inch-long key.

Set-Screws Used to Transmit Torque.—For certain applications it is common practice to use set-screws to transmit torque because they are relatively inexpensive to install and permit axial adjustment of the member mounted on the shaft. However, set-screws depend primarily on friction and the shearing force at the point of the screw, so they are not especially well-suited for high torques or where sudden load changes take place.

One rule for determining the proper size of a set-screw states that the diameter of the screw should equal $^{5}/_{16}$ inch plus one-eighth the shaft diameter. The holding power of set-screws selected by this rule can be checked using the formula on page **1834** of the *Handbook*.

PRACTICE EXERCISES FOR SECTION 17

(See *Answers to Practice Exercises for Section 17* on page **248**)

1) What is the polar section modulus of a 2-inch diameter shaft?

2) Using the information in the note at the bottom of page **3242** in the *Machinery's Handbook 32 Digital Edition*, redo #1 above using the tables *Section Moduli and Moments of Inertia for Round Shafts* starting on page **3242**.

3) If a 3-inch shaft is subjected to a torsional or twisting moment of 32,800 pound-inches, what is the equivalent torsional or shearing stress?

4) Is the shaft referred to in #2 above subjected to an excessive torsional stress?

5) If a 10-horsepower motor operating at its rated capacity connects by a belt with a 16-inch pulley on the driving shaft of a machine, what is the load tangential to the pulley rim and the

resulting twisting moment on the shaft, assuming that the rim speed of the driven pulley is 600 feet per minute?

6) How is the maximum distance between bearings for steel line-shafting determined?

7) What are "gib-head" keys, and why are they used on some classes of work?

8) What is the distinctive feature of Woodruff keys?

9) What are the advantages of Woodruff keys?

10) If a $3/8$-inch wide keyseat is to be milled into a $1\frac{1}{2}$-inch diameter shaft and if the keyseat depth is $3/16$ inch (as measured at one side), what is the depth from the top surface of the shaft or the amount to sink the cutter after it grazes the top of the shaft?

SECTION 18

SPLINES

Machinery's Handbook pages **2333–2362**

This section of the *Handbook* shows how to calculate the dimensions of involute splines and how to provide specifications for manufacturing drawings. Many types of mechanical connections between shafts and hubs are available for both fixed and sliding applications. Among these connections are the ordinary key and keyway (*Handbook* pages **2545–2568**), multiple keys and keyways, three- and four-lobed polygon shaft and hub connections, and involute splines of both inch dimension and metric module sizes.

The major advantages of involute splines are that they may be manufactured on the same equipment used to manufacture gears, they may be used for fixed and interference fit connections as well as for sliding connections, and they are stronger than most other connections with the exception of polygon-shaped members.

The section in the *Handbook* on involute splines, pages **2333–2359**, provides tables, data, formulas, and diagrams for American Standard splines made to both inch and metric module systems. Both systems share common definitions of terms, although the symbols used to identify dimensions and angles may differ, as shown on *Handbook* page **2354**. The two systems do not provide for interchangeability of parts; the new metric module Standard is the American National Standards Institute version of the International Standards Organization involute spline Standard, which is based upon metric, not inch, dimensions.

Example 1: A metric module involute spline pair is required to meet the following specification: pressure angle $\alpha_D = 30°$; module $m = 5$; number of teeth $Z = 32$; fit class = H/h; tolerance class 5 for both the internal and external splines; flat-root design for both members; length of engagement of the splines is 100 mm.

Table 13 beginning on *Handbook* page **2356** provides all the formulas necessary to calculate the dimensions of these splines.

Pitch diameter:

$$D = mZ = 5 \times 32 = 160 \ \text{mm} \tag{1}$$

Base diameter:

$$\begin{aligned} DB &= mZ\cos\alpha_D = 160 \times \cos\alpha_D = 160 \times \cos 30° \\ &= 160 \times 0.86603 = 138.5641 \ \text{mm} \end{aligned} \tag{2}$$

Circular pitch:

$$p = \pi m = 3.1416 \times 5 = 15.708 \tag{3}$$

Base pitch:

$$p_b = \pi m \cos\alpha_D = \pi \times 5 \times 0.86603 = 13.60350 \tag{4}$$

Tooth thickness modification:

$$es = 0 \tag{5}$$

in accordance with footnote to **Table 14**, *Handbook* page **2357**, and the Fit Classes paragraph on page **2354** that refers to H/h fits.

Minimum major diameter, internal spline,

$$DEI \text{ min} = m(Z + 1.8) = 5 \times (32 + 1.8) = 169.000 \tag{6}$$

Maximum major diameter, internal spline,

$$\begin{aligned} DEI \text{ max} &= DEI \text{ min} + (T + \lambda)/(\tan\alpha_D) \\ &= 169.000 + 0.248/\tan 30° \\ &= 169.4295 \ \text{mm} \end{aligned} \tag{7}$$

In this last calculation, the value of $(T + \lambda) = 0.248$ for class 7 was calculated using the formula in **Table 15**, *Handbook* page **2357**, as follows:

$$\begin{aligned} \text{i*} &= 0.001(0.45\sqrt[3]{D} + 0.001D) \\ &= 0.001(0.45\sqrt[3]{160} + 0.001 \times 160) \\ &= 0.00260 \end{aligned} \tag{8a}$$

$$\begin{aligned} \text{i**} &= 0.001(0.45\sqrt[3]{7.85398} + 0.001 \times 7.85398) \\ &= 0.00090 \end{aligned} \tag{8b}$$

In **Equation (8b)**, 7.85398 is the value of S_{bsc} calculated from the formula $S_{bsc} = 0.5\pi m$ given in the table starting on *Handbook* page **2356**.

$$
\begin{aligned}
(T + \lambda) &= 40i^* + 160i^* \\
&= 40 \times 0.00260 + 160 \times 0.00090 \\
&= 0.248 \text{ mm}
\end{aligned}
\qquad (8c)
$$

Form diameter, internal spline,

$$
\begin{aligned}
DFI &= m(Z + 1) + 2c_F \\
&= 5(32 + 1) + 2 \times 0.1m \\
&= 5(32 + 1) + 2 \times 0.1 \times 5 \\
&= 166 \text{ mm}
\end{aligned}
\qquad (9)
$$

In the above calculation the value of $c_F = 0.1m$ is taken from the diagram on *Handbook* page **2358**, and the corresponding formula for form clearance on *Handbook* page **2356**. Minimum minor diameter, internal spline,

$$
\begin{aligned}
DII \text{ min} &= DFE + 2c_F \\
&= 154.3502 + 2 \times 0.1 \times 5 \\
&= 155.3502 \text{ mm}
\end{aligned}
\qquad (10)
$$

The *DFE* value of 154.3502 used in this calculation was calculated from the formula on *Handbook* page **2356** as follows: $DB = 138.564$ from step (**2**); $D = 160$ from step (**1**); $h_s = 0.6m = 3.0$ from the last formula in the table starting on *Handbook* page **2356**; $es = 0$ from step (**5**); sin 30° = 0.50000; tan 30° = 0.57735. Therefore,

$$
DFE = 2 \times \sqrt{(0.5 \times 138.564)^2 + \left[0.5 \times 160 \times 0.50000 - \frac{0.6 \times 5 + \left(\frac{0.5 \times 0}{0.57735}\right)}{0.50000}\right]^2} \qquad (11)
$$

$$= 154.3502$$

Maximum minor diameter, internal spline,

$$DII\ \max = DII\ \min + (0.2m^{0.667} - 0.1m^{-0.5})$$
$$= 155.3502 + 0.58 \quad (12)$$
$$= 155.9302\ \text{mm}$$

The value 0.58 used in this calculation comes from the footnote *c* to the table on *Handbook* page **2356**. Circular space width, basic,

$$E_{bsc} = 0.5\pi m = 0.5 \times 3.1416 \times 5 = 7.854\ \text{mm} \quad (13)$$

Circular space width, minimum effective,

$$EV\ \min = E_{bsc} = 7.854\ \text{mm} \quad (14)$$

Circular space width, maximum actual,

$$E\ \max = EV\ \min + (T + \lambda)$$
$$= 7.854 + 0.0992 \quad \text{from step } (\mathbf{16c}) \quad (15)$$
$$= 7.9532\ \text{mm}$$

The value of $(T + \lambda)$ calculated in step **(16c)** is based upon class 5 fit stated at the beginning of the example. The value calculated in step **(8c)**, on the other hand, is based upon class 7 fit as required by the formula in step **(7)**. For class 5 fit, using the formula given in **Table 15**, *Handbook* page **2357**:

$$i^* = 0.00260 \quad \text{from step } (\mathbf{8a}) \quad (16a)$$

$$i^{**} = 0.00090 \quad \text{from step } (\mathbf{8b}) \quad (16b)$$

$$(T + \lambda) = 16i^* + 64i^{**} = 16 \times 0.00260 + 64 \times 0.00090$$
$$= 0.0992\ \text{mm} \quad (16c)$$

Circular space width, minimum actual,

$$E\ \min = EV\ \min + \lambda = 7.854 + 0.045 = 7.899\ \text{mm} \quad (17)$$

The value of λ used in this formula was calculated from the formulas for class 5 fit in **Table 16**, on *Handbook* page **2358**, and the formula in the text on *Handbook* page **2357** as follows:

$$F_p = 0.001(3.55\sqrt{5 \times 32 \times 3.1416/2} + 9) = 0.065 \text{ mm} \quad (18a)$$

$$f_f = 0.001[2.5 \times 5(1 + 0.0125 \times 32) + 16] = 0.034 \text{ mm} \quad (18b)$$

$$F_\beta = 0.001(1 \times \sqrt{100} \times 5) = 0.015 \text{ mm} \quad (18c)$$

$$\lambda = 0.6\sqrt{(0.065)^2 + (0.034)^2 + (0.015)^2} = 0.045 \text{ mm} \quad (18d)$$

Circular space width, maximum effective,

$$\begin{aligned} EV \max &= E \max - \lambda \qquad \textbf{(15)} \qquad \textbf{(18d)} \\ &= 7.9532 \text{ from step (15)} - 0.045 \text{ from step (18d)} \quad (19) \\ &= 7.9082 \text{ mm} \end{aligned}$$

Maximum major diameter, external spline,

$$\begin{aligned} DEE \max &= m(Z+1) - es/\tan\alpha_D = 5(32+1) - 0 \\ &= 165 \text{ mm} \end{aligned} \quad (20)$$

The value 0 in this last calculation is from **Table 17**, *Handbook* page **2358**, for h class fit.

Minimum major diameter, external spline, is calculated using the results of step **(20)** and footnote **c** on *Handbook* page **2356**,

$$\begin{aligned} DEE \min &= DEE \max - (0.2m^{0.667} - 0.01m^{-0.5}) \\ &= 165 - 0.58 = 164.42 \text{ mm} \end{aligned} \quad (21)$$

Maximum minor diameter, external spline,

$$\begin{aligned} DIE \max &= m(Z - 1.8) - es/\tan\alpha_D \\ &= 5(32 - 1.8) - 0 \\ &= 151 \text{ mm} \end{aligned} \quad (22)$$

The value 0 in this calculation is from **Table 17**, *Handbook* page **2358**, for h class fit.

Minimum minor diameter, external spline, is calculated using the results of steps (**22**) and (**7**),

$$\begin{aligned} DIE \text{ min} &= DIE \text{ max} - (T + \lambda)/\tan\alpha_D \\ &= 151 - 0.248/\tan 30° \\ &= 151 - 0.4295 \\ &= 150.570 \text{ mm} \end{aligned} \quad (23)$$

Circular tooth thickness, basic, has been taken from step (**13**),

$$S_{bsc} = 7.854 \text{ mm} \quad (24)$$

Circular tooth thickness, maximum effective, is calculated using the results of steps (**13**) and (**5**),

$$\begin{aligned} SV \text{ max} &= S_{bsc} - es \\ &= 7.854 - 0 \\ &= 7.854 \text{ mm} \end{aligned} \quad (25)$$

Circular tooth thickness, minimum actual, is calculated using the results of steps (**25**) and (**16c**),

$$S \text{ min} = SV \text{ max} - (T + \lambda) = 7.854 - 0.0992 = 7.7548 \text{ mm} \quad (26)$$

Circular tooth thickness, maximum actual, is calculated using the results of steps (**25**) and (**18d**),

$$\begin{aligned} S \text{ max} &= SV \text{ max} - \lambda \\ &= 7.854 - 0.045 \\ &= 7.809 \text{ mm} \end{aligned} \quad (27)$$

Circular tooth thickness, minimum effective, is calculated using the results of steps (**26**) and (**18d**),

$$\begin{aligned} SV \text{ min} &= S \text{ min} + \lambda \\ &= 7.754 + 0.045 \\ &= 7.799 \text{ mm} \end{aligned} \quad (28)$$

Example 2: As explained on *Handbook* page **2351**, spline gages are used for routine inspection of production parts. However, as part of an analytical procedure to evaluate effective space width or effective tooth thickness, measurements with pins are often used. Measurements with pins are also used for checking the actual space width and tooth thickness of splines during the machining process. Such measurements help in making the necessary size adjustments both during the setup process and as manufacturing proceeds. For the splines calculated in **Example 1**, what are the pin measurements for the tooth thickness and space width?

The maximum space width for the internal spline is 7.953 mm from step **(15)** in **Example 1**. The minimum tooth thickness for the external spline is 7.755 mm from step **(26)**.

Handbook page **2352** gives a method for calculating pin measurements for splines. This procedure was developed for inch-dimension splines. However, it may be used for metric module splines simply by replacing P wherever it appears in a formula by $1/m$; and by using millimeters instead of inches as dimensional units throughout.

For two-pin measurement *between* pins for the *internal* spline, steps **1)**, **2)**, and **3)** starting on *Handbook* page **2352** are used as follows:

$$\begin{aligned} \text{inv } \phi_i &= 7.953/160 + \text{inv } 30° - 8.64/138.564 \\ &= 0.049706 + 0.053751 - 0.062354 = 0.041103 \end{aligned} \quad (1)$$

The numbers used in this calculation are taken from the results in **Example 1** except for the involute of 30°, which is from the table on page **108** of the *Handbook*, and 8.64 is the diameter of the wire as calculated from the formula on *Handbook* page **2353**, $1.7280/P$ in which $1/m$ has been substituted for P to give $1.7280m = 1.7280 \times 5 = 8.64$. Note that the symbols on page **2352** are not the same as those used in **Example 1**. This is because the metric Standard for involute splines uses different symbols for the same dimensions. The table on page **2355** of the *Handbook* shows how these different symbols compare.

The value of inv $\phi_i = 0.041103$ is used to enter the table on *Handbook* page **108** to find, by interpolation,

$$\phi_i = 27°36'20'' \quad (2)$$

From a calculator find

$$\sec 27°36'20'' = 1.1285 \quad (3)$$

Calculate the measurement between wires:

$$\begin{aligned} M_i &= D_b \sec\phi_i - d_i = 138.564 \times 1.1285 - 8.64 \\ &= 147.729 \text{ mm} \end{aligned} \quad (4)$$

For two-pin measurement *over* the teeth of *external* splines, steps **1)**, **2)**, and **3)** on *Handbook* pages **2352–2353** are used as follows:

$$\text{inv } \phi_e = 7.755/160 + 0.053751 + 9.6/138.564 - 3.1416/32 = 0.073327 \quad (5)$$

Therefore, from *Handbook* page **109**, $\phi_e = 32°59'$ and, from a calculator, sec $32°59' = 1.1921$. From the formula in step **3)** on *Handbook* page **2353**:

$$M_e = 138.564 \times 1.1921 + 9.6 = 174.782 \text{ mm} \quad (6)$$

The pin diameter 9.6 in this calculation was calculated from the formula in step **3)** on *Handbook* page **2353** by substituting $1/m$ for P in the formula $d_e = 1.9200/P = 1.9200m$.

Specifying Spline Data on Drawings.—As stated on *Handbook* page **2346**, if the data specified on a spline drawing are suitably arranged and presented in a consistent manner, it is usually not necessary to provide a graphic illustration of the spline teeth. **Table 6** on *Handbook* page **2345** illustrates a flat-root spline similar to the one in **Example 1** except that it is an inch-dimension spline. The method of presenting drawing data for metric module splines differs somewhat from that shown on page **2345** in that the number of decimal places used for metric spline data is sometimes less than that for the corresponding inch-dimension system.

Example 3: How much of the data calculated or given in **Example 1** and **Example 2** should be presented on the spline drawing?

For the internal spline, the data required to manufacture the spline should be presented as follows, including the number of decimal places shown:

Example of Internal Involute Spline Drawing Data

Flat-Root Side Fit	Tolerance Class 5H
Number of Teeth	32
Module	5
Pressure Angle	30 deg
Base Diameter	138.5641 REF

Example of Internal Involute Spline Drawing Data *(Continued)*

Flat-Root Side Fit	Tolerance Class 5H
Pitch Diameter	160.0000 REF
Major Diameter	169.42 Max
Form Diameter	166.00
Minor Diameter	155.35/155.93
Circular Space Width:	
Max. Actual	7.953
Min. Effective	7.854
Max. Measurement Between Pins:	147.729 REF
Pin Diameter	8.640

For the external spline, the data required to manufacture the spline should be presented as follows:

Example of External Involute Spline Drawing Data

Flat-Root Side Fit	Tolerance Class 5H
Number of Teeth	32
Module	5
Pressure Angle	30 deg
Base Diameter	138.5641 REF
Pitch Diameter	160.0000 REF
Major Diameter	164.42/165.00
Form Diameter	154.35
Minor Diameter	150.57 MIN
Circular Space Width:	
Max. Actual	7.854
Min. Effective	7.809
Min. Measurement Over Pins:	74.782 REF
Pin Diameter	9.6

PRACTICE EXERCISES FOR SECTION 18

(See *Answers to Practice Exercises for Section 18* on page **249**)

1) What is the difference between a "soft" conversion of a standard and a "hard" system?

2) The Standard for metric module splines does not include a major diameter fit. What Standard provides for a major diameter fit?

3) What is an involute serration and is it still called this in American Standards?

4) What are some of the advantages of involute splines?

5) What is the meaning of the term "effective tooth thickness"?

6) What advantage is there in using an odd number of spline teeth?

7) If a spline connection is misaligned, fretting can occur at certain combinations of torque, speed, and misalignment angle. Is there any method for diminishing such damage?

8) For a given design of spline is there a method for estimating the torque capacity based upon wear? Based on shearing stress?

9) What does REF following a dimension of a spline mean?

10) Why are fillet-root splines sometimes preferred over flat-root splines?

SECTION 19

DESIGNING AND CUTTING GEARS

Machinery's Handbook pages **2206–2332**

In the design of gearing, there may be three distinct types of problems. These are: (1) determining the relative sizes of two or more gears to obtain a given speed or series of speeds; (2) determining the pitch of the gear teeth so that they will be strong enough to transmit a given amount of power; and (3) calculating the dimensions of a gear of a given pitch, such as the outside diameter, the depth of the teeth, and other dimensions needed in cutting the gear.

When the term "diameter" is applied to a spur gear, the pitch diameter is generally referred to and not the outside diameter. In calculating the speeds of gearing, the pitch diameters are used and not the outside diameters, because when gears are in mesh, the imaginary pitch circles roll in contact with each other.

Calculating Gear Speeds.—The simple rules for calculating the speeds of pulleys beginning on *Handbook* page **2569** may be applied to gearing, provided either the pitch diameters of the gears or the numbers of teeth are substituted for the pulley diameters. Information on gear speeds, especially as applied to compound trains of gearing, also will be found in the section dealing with lathe change gears beginning on *Handbook* page **2183**. When gear trains must be designed to secure unusual or fractional gear ratios, the directions beginning on *Handbook* page **2184** will be found very useful. A practical application of these methods is shown by examples beginning on *Handbook* page **2188**.

Planetary or epicyclic gearing is an increasingly important class of power transmission in various industries because of compactness, efficiency, and versatility. The rules for calculating rotational speeds and ratios are different from those for other types of gearing. Formulas for the most commonly used types of planetary gears are provided on *Handbook* pages **2293–2295**.

Example 1: The following example illustrates the method of calculating the speed of a driven shaft in a combination belt and gear drive when the diameters of the pulleys and the pitch diameters of the gears are known, and the number of revolutions per minute (rpm) of the driving shaft is given. If driving pulley A, **Fig. 1**, is 16 inches in diameter, and driven pulley B, 6 inches in diameter, and the pitch diameter of driving gear C is 12 inches, driving gear D is 14 inches, driven gear E, 7 inches, driven gear F, 6 inches, and driving pulley A makes 60 rpm, determine the number of rpm of F.

$$\frac{16 \times 12 \times 14}{6 \times 7 \times 6} \times 60 = 640 \text{ rpm}$$

The calculations required in solving problems of this kind can be simplified if the gears are considered as pulleys having diameters equal to their pitch diameters. When this is done, the rules that apply to compound belt drives can be used in determining the speed or size of the gears or pulleys.

Substituting the numbers of teeth in each gear for the pitch diameter gives the same result as when the pitch diameters are used.

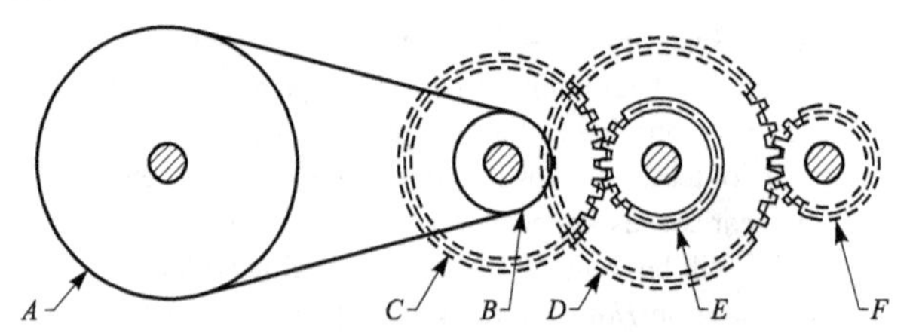

Fig. 1. Combination Pulley and Compound Gear Drive

Example 2: If driving spur gear A (**Fig. 2**) makes 336 rpm and has 42 teeth, driven spur gear B, 21 teeth, driving bevel gear C, 33 teeth, driven bevel gear D, 24 teeth, driving worm E, one thread, and driven worm-wheel F, 42 teeth, determine the number of rpm of F.

When a combination of spur, bevel, and wormgearing is employed to transmit motion and power from one shaft to another, the speed of the driven shaft can be found by the following method: Consider the worm as a gear having one tooth if it is single-threaded and as a gear having two teeth if double-threaded, etc. The speed of the driving shaft can then be found by applying the rules for ordinary compound spur gearing. In this example,

$$\frac{42 \times 33 \times 1}{21 \times 24 \times 42} \times 336 = 22 \text{ rpm}$$

If the pitch diameters of the gears are used instead of the number of teeth in making calculations, the worm should be considered as a gear having a pitch diameter of 1 inch if single-threaded, and 2 inches if a double-threaded worm, etc.

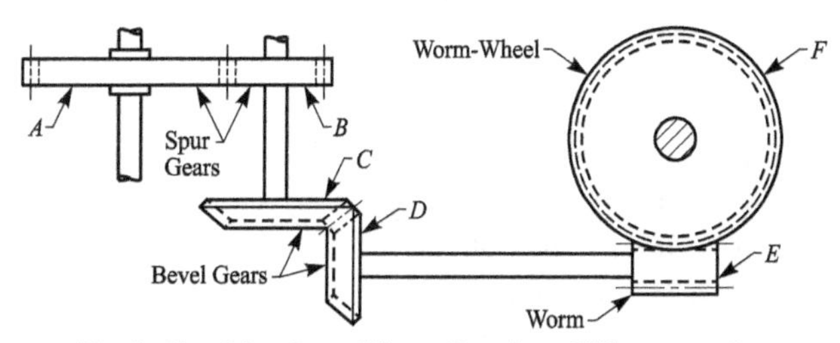

Fig. 2. Combination of Spur, Bevel, and Wormgearing

Example 3: If a worm is triple-threaded and makes 180 rpm, and the worm-wheel is required to make 5 rpm, determine the number of teeth in the worm-wheel.

Rule: Multiply the number of threads in the worm by its number of rpm, and divide the product by the number of rpm of the worm-wheel. By applying this rule,

$$\frac{3 \times 180}{5} = 108 \text{ teeth}$$

Example 4: A 6-inch grinding machine with a spindle speed of 1773 rpm, for a recommended peripheral speed of 6500 feet per minute (as figured for a full-size 14-inch wheel for this size of machine), has two steps on the spindle pulley; the large step is 5.5 inches in diameter and the small step, 4 inches. What should be the minimum diameter of the wheel before the belt is shifted to the smaller step in order to select a peripheral wheel speed of 6500 feet per minute?

As the spindle makes 1773 rpm when the belt is on the large pulley, its speed with the belt on the smaller pulley may be determined as follows: 5.5:4 = x:1773, or (5.5 × 1773)/4 = 2438 rpm, approximately. To obtain the same peripheral speed as when the belt is on the large pulley, the diameters of the grinding wheel should be 14:x = 2438:1773, or (14 × 1773)/2438 = 10.18 inches. Therefore, when the grinding wheel has been worn down to a diameter of 10.18 inches, or approximately 10$^3/_{16}$ inches, the spindle belt should be shifted to the smaller step of the spindle pulley to obtain a peripheral speed of 6500 feet per minute. The method used in this example may be reduced to a formula for use with any make of grinding machine having a two-step spindle pulley.

Let

D = diameter of wheel, full size

D_1 = diameter of wheel, reduced size

d = diameter of large pulley step

d_1 = diameter of small pulley step

V = spindle rpm, using large pulley step

v = spindle rpm, using small pulley step

Then,

$$v = \frac{dV}{d_1};\quad D_1 = \frac{DV}{v}$$

Example 5: Planetary gear sets are widely used in power transmission because of their compactness and relatively high efficiency when properly designed. The simple planetary configuration shown in **Fig. 10** on *Handbook* page **2294** is typical of high-efficiency designs. If $A = 20$ and $C = 40$, what is the rotation of the driver D per revolution of the follower?

Using the formula given on *Handbook* page **2294**,

$$D = 1 + \frac{C}{A} = 1 + \frac{40}{20} = 3$$

Example 6: If, in **Example 5**, the diameter of the fixed gear is doubled to $C = 80$, what effect does that produce in the rotation of the drive D?

$$D = 1 + \frac{80}{20} = 5$$

Note that doubling the size of the fixed gear C does not double the ratio or the driver speed of the gear set because the overall ratio is always plus the ratio of C to A.

Example 7: The compound type of planetary gear shown in **Fig. 13** on *Handbook* page **2294** can provide high revolution ratios, although the efficiency decreases as the ratio increases. What is the rotation of the follower F when $B = 61$, $C = 60$, $x = 19$, and $y = 20$?

$$F = 1 - \left(\frac{C \times x}{y \times B}\right) = 1 - \left(\frac{60 \times 19}{20 \times 61}\right) = 1 - \frac{57}{61} = 0.06557$$

Example 8: In **Example 7**, what is the rotation of the driver per revolution of the follower?

$$\text{Driver} = \frac{1}{\text{follower}} = \frac{1}{0.06557} = 15.25$$

Note that in compound planetary gear drives the sum of meshing tooth pairs must be equal for proper meshing. Thus, $C + y = x + B$.

Diametral Pitch of a Gear.—The diametral pitch represents the number of gear teeth for each inch of pitch diameter and, therefore, equals the number of teeth divided by the pitch diameter. The term "diametral pitch" as applied to bevel gears has the same meaning as with spur gears. This method of basing the pitch on the relation between the number of teeth and the pitch diameter is used almost exclusively in connection with cut gearing and to some extent for cast gearing. The term "circular pitch," which refers to the distance between the centers of adjacent teeth measured along the pitch circle, is used for cast gearing but very little for cut gearing except very large sizes. If 3.1416 is divided by the diametral pitch, the quotient equals the circular pitch, or, if the circular pitch is known, the diametral pitch may be found by dividing 3.1416 by the circular pitch. The pitch of the gear teeth may depend primarily upon the strength required to transmit a given amount of power.

Power-Transmitting Capacity of Bevel Gears.—The design of bevel gears to meet a set of operating conditions is best accomplished in four steps: (1) determine the design load upon which the bevel gear sizes will be based; (2) using design literature and charts available from gear manufacturers and distributors, select approximate gear and pinion sizes to satisfy the load requirements; (3) determine the maximum safe tooth load, based on gear geometry and material, using manufacturer's and/or AGMA formulas; and (4) determine the safe horsepower capacity of the gears, based on safe tooth load and tooth surface durability. The horsepower capacity of the gears should meet or exceed the design load requirements. To check the capacity of an existing bevel-gear drive, only steps (3) and (4) are necessary.

Dimensions and Angles Required in Producing Gears.—Many of the rules and formulas given in the gear section of the *Handbook* beginning on page **2206** are used in determining tooth dimensions, gear-blank sizes, and angles in bevel, helical, and wormgearing. These dimensions or angles are required on the working drawings used in connection with machining operations, such as turning gear blanks and cutting the teeth.

Example 9: If a spur gear is to have 40 teeth of 8 diametral pitch, to what diameter should the blank be turned? By applying **Example (4a)**, *Handbook* page **2212**, (40 + 2)/8 = 5.25 inches. Therefore, the outside diameter of this gear or the diameter to which the blank would be turned is $5\frac{1}{4}$ inches.

For internal spur gears, the inside diameter to which the gear blank would be bored may be obtained by subtracting 2 from the number of teeth and dividing the remainder by the diametral pitch.

Example 10: A sample spur gear has 22 teeth, and the outside diameter, or diameter measured across the tops of the teeth, is 6 inches. Determine the diametral pitch. According to **Example (4a)**, *Handbook* page **2212**,

$$D_o = \frac{N+2}{P}$$

Hence,

$$P = \frac{N+2}{D_o} = \frac{22+2}{6} = 4 \text{ diametral pitch}$$

The table, *Handbook* page **2212**, also shows that when the sample gear has American Standard Stub teeth, **Example (6a)** should be used to determine the outside diameter, or diametral pitch.

Example 11: A 25-degree involute full-depth spur gear is to be produced by hobbing. How is the hob tip radius found?

Per *Handbook* pages **2237–2238**, the maximum hob tip radius, r_c (max), is found by the formula:

$$r_c\ (\max) = \frac{0.785398 \cos\phi - b \sin\phi}{1 - \sin\phi}$$

where ϕ is the pressure angle, here, 25°, and b is the dedendum constant, which is 1.250 according to **Table 2** on *Handbook* page **2212**. Thus,

$$r_c\ (\max) = \frac{0.785398 \times 0.90631 - 1.25 \times 0.42262}{1 - 0.42262}$$

$$= 0.3179 \text{ inch for a 1-diametral-pitch gear}$$

Example 12: If a 20-degree involute full-depth pinion having 24 teeth of 6 diametral pitch is to mesh with a rack, determine the whole depth of the rack teeth and the linear pitch of the teeth.

The teeth of a rack are of the same proportions as the teeth of a spur gear or pinion that is intended to mesh with the rack; hence the pitch of the rack teeth is equal to the circular pitch of the pinion and is found by dividing 3.1416 by the diametral pitch.

The pitch = 3.1416 ÷ 6 = 0.5236 inch = linear pitch of a rack to mesh with a pinion of 6 diametral pitch. This dimension (0.5236) represents the distance that the cutter would be indexed when milling rack teeth or the distance that the planer tool would be moved for cutting successive teeth if a planer were used. The whole depth of a full-depth rack tooth of 20-degree pressure angle equals 2.157 divided by the diametral pitch of the meshing gear, or the whole depth equals the circular pitch multiplied by 0.6866. Here, the circular pitch is 0.5236, and the whole depth equals 0.5236 × 0.6866 = 0.3595 inch.

Example 13: If the teeth of a spur gear are to be cut to a certain diametral pitch, is it possible to obtain any diameter that may be desired? Thus, if the diametral pitch is 4, is it possible to make the pitch diameter $5\frac{1}{8}$ inches?

The diametral pitch system is so arranged as to provide a series of tooth sizes, just as the pitches of screw threads are standardized. In as much as there must be a whole number of teeth in each gear, it is apparent that gears of a given pitch vary in diameter according to the number of teeth. Suppose, for example, that a series of gears is of 4 diametral pitch. Then the pitch diameter of a gear having, say, 20 teeth will be 5 inches; 21 teeth, $5\frac{1}{4}$ inches; 22 teeth, $5\frac{1}{2}$ inches, and so on. It will be seen that the increase in diameter for each additional tooth is equal to $\frac{1}{4}$ inch for 4 diametral pitch. Similarly, for 2 diametral pitch, the variations for successive numbers of teeth would equal $\frac{1}{2}$ inch, and for 10 diametral pitch the variations would equal $\frac{1}{10}$ inch, etc.

The center-to-center distance between two gears is equal to one-half the total number of teeth in the gears divided by the diametral pitch. It may be desirable at times to have a center distance that cannot be obtained exactly by any combination of gearing of given diametral pitch, but this condition is unusual, and, ordinarily, the designer of a machine can alter the center distance whatever slight amount may be required for gearing of the desired ratio and pitch. By using a standard system of pitches, all calculations are simplified, and it is also possible to obtain the benefits of standardization in the manufacturing of gears and gear-cutters.

Proportioning Spur Gears When Center Distance Is Fixed.—If the center-to-center distance between two shafts is fixed, and it is desired to use gears of a certain pitch, the number of teeth in each gear for a given speed may be determined as follows: Since the gears must be of a certain pitch, the total number of teeth available should be determined and then the number of teeth in the driving and the driven gears. The total number of teeth equals twice the product of the center distance multiplied by the diametral pitch. If the center distance is 6 inches, and the diametral pitch 2, the total number of teeth equals $6 \times 2 \times 10 = 120$ teeth. The next step is to find the number of teeth in the driving and the driven gears for a given rate of speed.

Rule: Divide the speed of the driving gear in revolutions per minute by the speed of the driven gear and add one to the quotient. Next divide the total number of teeth in both gears by the sum previously obtained, and the quotient will equal the number of teeth required in the driving gear. This number subtracted from the total number of teeth will equal the number of teeth required in the driven gear.

Example 14: If the center-to-center distance is 6 inches, and the diametral pitch is 10, the total number of teeth available will be 120. If the speeds of the driving and the driven gears are to be 100 and 60 revolutions per minute, respectively, find the number of teeth for each gear.

$^{100}/_{60} = 1^2/_3$ and $1^2/_3 + 1 = 2^2/_3$

$120 \div 2^2/_3 = {}^{120}/_1 \times {}^3/_8 = 45 =$ number of teeth in driving gear

The number of teeth in the driven gear equals 120 − 45 = 75 teeth.

When the center distance and the velocity ratios are fixed by some essential construction of a machine, it is often impossible to use standard-diametral-pitch gear teeth. If cast gears are to be used, it does not matter so much, as a pattern maker can lay out the teeth according to the pitch desired, but if cut gears are required, an effort should be made to alter the center distance so that standard-diametral-pitch cutters can be used since these are usually carried in stock.

Dimensions in Generated Bevel Gears.—*Example 15:* Find all the dimensions and angles necessary to manufacture a pair of straight bevel gears if the number of teeth in the pinion is 16, the number of teeth in the mating gear is 49, the diametral pitch is 5, and the face width is 1.5 inches. The gears are to have a 20-degree pressure angle, a 90-degree shaft angle, and must be in accordance with the Gleason System.

On page **186** of this *Guide*, **Table 1** gives formulas for Gleason System 20-degree pressure angle straight bevel gears with 90-degree shaft angle. These formulas are given in the same order as is normally used in computation. Computations of the gear dimensions should be arranged as shown in the table on the following pages to establish a consistent procedure when calculations for bevel gears are required frequently.

Given:

Number of pinion teeth, n	= 16	(1)
Number of gear teeth, N	= 49	(2)
Diametral pitch, P	= 5	(3)
Face width, F	= 1.5	(4)
Pressure angle, $\phi = 20°$	= 20°	(5)
Shaft angle, $\Sigma = 90°$	= 90°	(6)

Sample calculations for (**7**) to (**24**) are given in **Table 2** on page **188**.

Table 1. Formulas for Gleason System 20-Degree Straight Bevel Gears—90-Degree Shaft Angle

To Find		Formula	
No.	Item	Pinion	Gear
7	Working Depth	$h_k = \dfrac{2.000}{P}$	Same as pinion
8	Whole Depth	$h_t = \dfrac{2.188}{P} + 0.002$	Same as pinion
9	Pitch Diameter	$d = \dfrac{n}{P}$	$D = \dfrac{N}{P}$
10	Pitch Angle	$\gamma = \tan^{-1}\dfrac{n}{N}$	$\Gamma = 90° - \gamma$
11	Cone Distance	$A_O = \dfrac{D}{2\sin\Gamma}$	Same as pinion
12	Circular Pitch	$p = \dfrac{3.1416}{P}$	Same as pinion
13	Addendum	$a_p = h_t - a_G$	$a_G = \dfrac{0.540}{P} + \dfrac{0.460}{P\left(\dfrac{N}{n}\right)^2}$
14	Dedendum[a]	$b_p = \dfrac{2.188}{P} - a_p$	$b_G = \dfrac{2.188}{P} - a_G$
15	Clearance	$c = h_t - h_k$	Same as pinion
16	Dedendum Angle	$\delta_P = \tan^{-1}\dfrac{b_p}{A_O}$	$\delta_G = \tan^{-1}\dfrac{b_G}{A_O}$
17	Face Angle of Blank	$\gamma_O = \gamma + \delta_G$	$\Gamma_O = \Gamma + \delta_p$
18	Root Angle	$\gamma_r = \gamma - \delta_p$	$\Gamma_R = \Gamma - \delta_G$
19	Outside Diameter	$d_O = d + 2a_p\cos\gamma$	$D_O = D + 2a_G\cos\Gamma$

Table 1. *(Continued)* **Formulas for Gleason System 20-Degree Straight Bevel Gears—90-Degree Shaft Angle**

To Find			
		Formula	
No.	Item	Pinion	Gear
20	Pitch Apex to Crown	$x_O = \frac{D}{2} - a_p \sin\gamma$	$X_O = \frac{d}{2} - a_G \sin\Gamma$
21	Circular Thickness	$t = p - T$	$T = \frac{p}{2} - (a_p - a_G)\tan\phi - \frac{K}{P}$ $K = (\text{Chart } 1)$
22	Backlash	$B =$ (See table on *Handbook* page **2245**)	
23	Chordal Thickness	$t_c = t - \frac{t^3}{6d^2} - \frac{B}{2}$	$T_c = T - \frac{T^3}{6D^2} - \frac{B}{2}$
24	Chordal Addendum	$a_{cp} = a_p + \frac{t^2 \cos\gamma}{4d}$	$a_{CG} = a_G + \frac{T^2 \cos\gamma}{4D}$
25	Tooth Angle	$\frac{3438}{A_O}\left(\frac{t}{2} + b_p \tan\phi\right)$ minutes	$\frac{3438}{A_O}\left(\frac{T}{2} + b_G \tan\phi\right)$ minutes
26	Limit Point Width	$\frac{A_O - F}{A_O}(T - 2b_p \tan\phi) - 0.0015$	$\frac{A_O - F}{A_O}(t - 2b_G \tan\phi) - 0.0015$

[a] The actual dedendum will be 0.002 inches greater than calculated due to tool advance

All linear dimensions are in inches.

The tooth angle (Item 25, **Table 1**) is a machine setting and is only computed if a Gleason two-tool type straight bevel gear generator is to be used. Calculations continue on page **188**.

Dimensions of Milled Bevel Gears.—As explained on *Handbook* page **2262**, the tooth proportions of milled bevel gears differ in some respects from those of generated bevel gears. To take these differences into account, a separate table of formulas is given on *Handbook* page **2264** for use in calculating dimensions of milled bevel gears.

Example 16: Compute the dimensions and angles of a pair of mating bevel gears that are to be cut on a milling machine using rotary formed milling cutters if the data given are as follows:

Number of pinion teeth	=	15
Number of gear teeth	=	60
Diametral pitch	=	3
Face width	=	1.5
Pressure angle	=	$14\frac{1}{2}°$
Shaft angle	=	90°

Table 2. Calculations of Dimensions for Example 15

	Dimension	Pinion	Gear
(7)	Working depth	$2.000/5 = 0.400$	Same as Pinion
(8)	Whole depth	$2.188/5 + 0.002 = 0.440$	Same as Pinion
(9)	Pitch diameter	$^{16}/_{5} = 3.2000$	$^{49}/_{5} = 9.8000$
(10)	Pitch angle	$\tan^{-1}(^{16}/_{49}) = 18°\ 5'$	$90° - 18°5' = 71°55'$
(11)	Cone distance	$9.8000/(2 \times \sin 71°55') = 5.1546$	Same as pinion
(12)	Circular pitch	$3.1416 / 5 = 0.6283$	Same as pinion
(13)	Addendum	$0.400 - 0.118 = 0.282$	$0.540/5 + 0.460/(5(49/16)^2) = 0.118$
(14)	Dedendum	$2.188/5 - 0.282 = 0.1554$	$2.188/5 - 0.118 = 0.3196$
(15)	Clearance	$0.440 - 0.400 = 0.040$	Same as pinion
(16)	Dedendum angle	$\tan^{-1}(0.1536/5.1546) = 1°42'$	$\tan^{-1}(0.3214/5.1546) = 3°34'$
(17)	Face angle of blank	$18°5' + 3°34' = 21°39'$	$71°55' + 1°42' = 73°37'$
(18)	Root angle	$18°5' \angle 1°42' = 16°23'$	$71°55' \angle 3°34' = 68°21'$
(19)	Outside diameter	$3.2000 + 2 \times 0.282 \cos 18°5' = 3.735$	$9.8000 + 2 \times 0.118 \cos 71°55' = 9.875$
(20)	Pitch apex to crown	$9.8000/2 - 0.284 \sin 18°5' = 4.812$	$3.2000/2 - 0.118 \sin 71°55' = 1.488$
(21)	Circular thickness	$0.6283 - 0.2467 = 0.3816$	$0.6283/2 - (0.284 - 0.118)\tan 20° - (0.038(\text{chart } 1))/5 = 0.2467$
(22)	Backlash	0.006	0.006
(23)	Chordal thickness	$0.3816 - \dfrac{(0.3816)^3}{6 \times (3.2000)^2} - \dfrac{0.006}{2} = 0.378$	$0.2467 - \dfrac{(0.2467)^3}{6 \times (9.8000)^2} - \dfrac{0.006}{2} = 0.244$
(24)	Chordal addendum	$0.282 + \dfrac{0.3816^2 \cos 18°5'}{4 \times 3.2000} = 0.293$	$0.118 + \dfrac{0.2467^2 \cos 71°55'}{4 \times 9.8000} = 0.118$

By using the formulas on *Handbook* page **2264**,

$$\tan \alpha_p = 15 \div 60 = 0.25 = \tan 14° 2.2', \text{say}, 14° 2'$$

$$\alpha_G = 90° - 14° 2' = 75° 57.8', \text{say}, 75° 58'$$

$$D_p = 15 \div 3 = 5.0000 \text{ inches}$$

$$D_G = 60 \div 3 = 20.0000 \text{ inches}$$

$$S = 1 \div 3 = 0.3333 \text{ inch}$$

$$S + A = 1.157 \div 3 = 0.3857 \text{ inch}$$

$$W = 2.157 \div 3 = 0.7190 \text{ inch}$$

$$T = 1.571 \div 3 = 0.5236 \text{ inch}$$

$$C = \frac{5.000}{2 \times 0.24249} = 10.308 \text{ inches}$$

(In determining C, the sine of unrounded value of α_p, 14° 2.2′, is used.)

$$F = 8 \div 3 = 2\tfrac{2}{3}, \text{say}, 2\tfrac{5}{8} \text{ inches}$$

$$s = 0.3333 \times \frac{10.308 - 2\frac{5}{8}}{10.308} = 0.2484 \text{ inch}$$

$$t = 0.5236 \times \frac{10.308 - 2\frac{5}{8}}{10.308} = 0.3903 \text{ inch}$$

$$\tan\theta = 0.3333 \div 10.308 = \tan 1° 51'$$

$$\tan\phi = 0.3857 \div 10.308 = \tan 2° 9'$$

$$\gamma_P = 14° 2' + 1° 51' = 15° 53'$$

$$\gamma_G = 75° 58' + 1° 51' = 77° 49'$$

$$\delta_P = 90° - 15° 53' = 74° 7'$$

$$\delta_G = 90° - 77° 49' = 12° 11'$$

$$\zeta_P = 14° 2' + 2° 9' = 11° 53'$$

$$\zeta_G = 75° 582' + 2° 9' = 73° 49'$$

$$K_P = 0.3333 \times 0.97015 = 0.3234 \text{ inch}$$

$$K_G = 0.3333 \times 0.24249 = 0.0808 \text{ inch}$$

$$O_P = 5.000 + 2 \times 0.3234 = 5.6468 \text{ inches}$$

$$O_G = 20.000 + 2 \times 0.0808 = 20.1616 \text{ inches}$$

$$J_P = \frac{5.6468}{2} \times 3.5144 = 9.9226 \text{ inches}$$

$$J_G = \frac{20.1616}{2} \times 0.21590 = 2.1764 \text{ inches}$$

$$j_p = 9.9226 \times \frac{10.3097 - 2\tfrac{5}{8}}{10.3097} = 7.3961 \text{ inches}$$

$$j_g = 2.1764 \times \frac{10.3097 - 2\tfrac{5}{8}}{10.3097} = 1.6222 \text{ inches}$$

$$N'_P = \frac{15}{0.97015} = 15.4\text{, say, 15 teeth}$$

$$N'_G = \frac{60}{0.24249} = 247 \text{ teeth}$$

If these gears are to have uniform clearance at the bottom of the teeth, in accordance with the recommendation given in the last paragraph on *Handbook* page **2262**, then the cutting angles ζ_P and ζ_G should be determined by subtracting the addendum angle from the pitch cone angles. Thus,

$$\zeta_P = 14°2' - 1°51' = 12°11'$$

$$\zeta_G = 75°58' - 1°51' = 74°7'$$

Selection of Formed Cutters for Bevel Gears.—*Example 17:* In **Example 16**, the numbers of teeth for which to select the cutters were calculated as 15 and 247 for the pinion and gear, respectively. Therefore, as explained on page **2268** of the *Handbook*, the cutters selected from the table on page **2231** are the No. 7 and the No. 1 cutters. As further noted on page **2268**, bevel gear milling cutters may be selected directly from the table beginning on page **2266**, when the shaft angle is 90 degrees, instead of using the computed value of N' to enter the table on page **2231**. Thus, for a 15-tooth pinion and a 60-tooth gear, the table on page **2266** shows that the numbers of the cutters to use are 1 and 7 for gear and pinion, respectively.

Pitch of Hob for Helical Gears.—*Example 18:* A helical gear that is to be used for connecting shafts has 83 teeth, a helix angle of 7 degrees, and a pitch diameter of 47.78 inches. Determine the pitch of hob to use in cutting this gear.

As explained on *Handbook* page **2277**, the normal diametral pitch and the pitch of the hob are determined as follows: the transverse diametral pitch equals 83 ÷ 47.78 = 1.737. The cosine of the helix angle of the gear (7 degrees) is 0.99255; hence the normal diametral pitch equals 1.737 ÷ 0.99255 = 1.75; therefore, a hob of $1\frac{3}{4}$ diametral pitch should be used. This hob is the same as would be used for spur gears of $1\frac{3}{4}$ diametral pitch, and it will cut any spur or helical gear of that pitch regardless of the number of teeth, provided $1\frac{3}{4}$ is the diametral pitch of the spur gear and the normal diametral pitch of the helical gear.

Determining Contact Ratio.—As pointed out on *Handbook* page **2237**, if a smooth transfer of load is to be obtained from one pair of teeth to the next pair of teeth as two mating gears rotate under load, the contact ratio must be well over 1.0. Usually, this ratio should be 1.4 or more, although in extreme cases it may be as low as 1.15.

Example 19: Find the contact ratio for a pair of 18-diametral pitch, 20-degree pressure gears, one having 36 teeth and the other 90 teeth. From **Formula (1)** given on *Handbook* page **2236**:

$$\cos A = \frac{90 \times \cos 20°}{5.111 \times 18} = \frac{90 \times 0.93969}{91.9998} = 0.91926 \quad \text{and}$$

$$A = 23°11'$$

From **Formula (4)** given on *Handbook* page **2236**:

$$\cos a = \frac{36 \times \cos 20°}{2.111 \times 18} = \frac{36 \times 0.93969}{37.9998} = 0.89024 \quad \text{and}$$

$$a = 27°6'$$

From **Formula (5)** given on *Handbook* page **2237**:

$$\tan B = \tan 20° - \frac{36}{90}(\tan 27°6' - \tan 20°)$$

$$= 0.36397 - \frac{36}{90}(0.51172 - 0.36397) = 0.30487$$

From **Formula (7a)** given on *Handbook* page **2237**, the contact ratio m_f is found:

$$m_f = \frac{90}{6.28318}(0.42826 - 0.30487)$$

$$= 1.77$$

which is satisfactory.

Dimensions Required When Using Enlarged Fine-Pitch Pinions.— On *Handbook* pages **2232–2234**, there are tables of dimensions for enlarged fine-pitch pinions. These tables show how much the dimensions of enlarged pinions must differ from standard when the number of teeth is small, and undercutting of the teeth is to be avoided.

Example 20: If a 10- and a 31-tooth mating pinion and gear of 20 diametral pitch and $14\frac{1}{2}°$ pressure angle have both been enlarged to avoid undercutting of the teeth, what increase over the standard center distance is required?

$$\text{Standard center distance} = \frac{n + N}{2P} = \frac{10 + 31}{2 \times 20} = 1.0250 \text{ inches}$$

The amount by which the center distance must be increased over standard can be obtained by taking the sum of the amounts shown in the eighth column of **Table 9b** on *Handbook* page **2232** and dividing this sum by the diametral pitch. Thus, the increase over the standard center distance is (0.6866 + 0.0283)/20 = 0.0357 inch.

Example 21: At what center distance would the gears in **Example 20** have to be meshed if there were to be no backlash?

Obtaining the two thicknesses of both gears at the standard pitch diameters from **Table 9b** on *Handbook* page **2232**, dividing them by 20, and using the formulas on *Handbook* page **2236**:

$$\text{inv}\phi_1 = \text{inv}14\tfrac{1}{2}° + \frac{20(0.09630 + 0.07927) - 3.1416}{10 + 31}$$

The involute of $14\frac{1}{2}°$ is found on *Handbook* page **107** to be 0.0055448. Therefore,

$$\text{inv}\phi_1 = 0.0055448 + 0.0090195 = 0.0145643$$

By referring again to the table on *Handbook* page **107**:

$$\phi_1 = 19°51'6''$$

$$C = \frac{n+N}{2P} = \frac{10+31}{2 \times 20} = 1.025 \text{ inch}$$

$$C_1 = \frac{\cos 14\frac{1}{2}°}{\cos 19°51'6''} \times 1.025 = \frac{0.96815}{0.94057} \times 1.025 = 1.0551 \text{ inch}$$

End Thrust of Helical Gears Applied to Parallel Shafts.— *Example 22:* The diagrams on *Handbook* pages **2278–2279** show the application of helical or spiral gears to parallel shaft drives. If a force of 7 horsepower is to be transmitted at a pitch-line velocity of 200 feet per minute, determine the end thrust in pounds, assuming that the helix angle of the gear is 15 degrees.

To determine the end thrust of helical gearing as applied to parallel shafts, first calculate the tangential load on the gear teeth.

$$\text{Tangential load} = \frac{33{,}000 \times 7}{200} = 1155 \text{ pounds}$$

(This formula is derived from the formulas for power given on *Handbook* page **187**.)

The axial or end thrust may now be determined approximately by multiplying the tangential load by the tangent of the tooth angle. Thus, in this instance, the thrust = 1155 × tan 15 degrees = about 310 pounds. (Note that this formula agrees with the one on *Handbook* page **170** for determining force *P* parallel to base of inclined plane.) The end thrust obtained by this calculation will be somewhat greater than the actual end thrust, because frictional losses in the shaft bearings, etc., have not been taken into account, although a test on a helical gear set, with a motor drive, showed that the actual thrust of the $7\frac{1}{2}$-degree helical gears tested was not much below the values calculated as just explained.

According to most textbooks, the maximum angle for single helical gears should be about 20 degrees, although one prominent manufacturer mentions that the maximum angle for industrial drives ordinarily does not exceed 10 degrees, and this will give quiet running without excessive end thrust. On some of the heavier single helical gearing used for street railway transmissions, etc., an angle of 7 degrees is employed.

Dimensions of Wormgear Blank and the Gashing Angle.— *Example 23:* A wormgear having 45 teeth is to be driven by a double-threaded worm having an outside diameter of $2\frac{1}{2}$ inches and a lead of 1 inch, the linear pitch being $\frac{1}{2}$ inch. The throat diameter and throat radius of the wormgear are required as well as the angle for gashing the blank.

The throat diameter D_t equals the pitch diameter D plus twice the addendum A; thus, $D_t = D + 2A$. The addendum of the worm thread equals the linear pitch multiplied by 0.3183, and here, $0.5 \times 0.3183 = 0.1591$ inch. The pitch diameter of the wormgear $= 45 \times 0.5 \div 3.1416 = 7.162$ inches; hence, the throat diameter equals $7.162 + 2 \times 0.1591 = 7.48$ inches.

The radius of the wormgear throat is found by subtracting twice the addendum of the worm thread from $\frac{1}{2}$ the outside diameter of the worm. The addendum of the worm thread equals 0.1591 inch, and the radius of the throat, therefore, equals $(2.5 \div 2) - 2 \times 0.1591 = 0.931$ inch.

When a wormgear is hobbed in a milling machine, gashes are milled before the hobbing operation. The table must be swiveled around while gashing, the amount depending upon the relation between the lead of the worm thread and the pitch circumference. The first step is to find the circumference of the pitch circle of the worm. The pitch diameter equals the outside diameter minus twice the addendum of the worm thread; hence, the pitch diameter equals $2.5 - 2 \times 0.1591 = 2.18$ inches, and the pitch circumference equals $2.18 \times 3.1416 = 6.848$ inches.

Next, divide the lead of the worm thread by the pitch circumference to obtain the tangent of the desired angle, and then refer to a table of tangents or a calculator to determine what this angle is. For this example, it is $1 \div 6.848 = 0.1460$, which is the tangent of $8\frac{1}{3}$ degrees from its normal position.

Change Gear Ratio for Diametral-Pitch Worms.— *Example 24:* In cutting worms to a given diametral pitch, the ratio of the change gears is $22 \times$ threads per inch / $7 \times$ diametral pitch.

The reason why the constants 22 and 7 are used in determining the ratio of change-gears for cutting worm threads is because $\frac{22}{7}$

equals, very nearly, 3.1416, which is the circular pitch equivalent to diametral pitch.

Assume that the diametral pitch of the wormgear is 5, and the lathe screw constant is 4. (See *Handbook* page **2183** for the meaning of "lathe screw constant.") Then, (4 × 22)/(5 × 7) = 88/35. If this simple combination of gearing were used, the gear on the stud would have 88 teeth and the gear on the lead screw, 35 teeth. Of course, any other combination of gearing having this same ratio could be used, as, for example, the following compound train of gearing: (24 × 66)/(30 × 21).

If the lathe screw constant is 4, as previously assumed, then the number of threads per inch obtained with gearing having a ratio of 88/35 = (4 × 35)/88 = 1.5909; hence, the pitch of the worm thread equals 1 ÷ 1.5909 = 0.6284 inch, which is the circular pitch equivalent to 5 diametral pitch, correct to within 0.0001 inch.

Bearing Loads Produced by Bevel Gears.—In applications where bevel gears are used, not only must the gears be proportioned with regard to the power to be transmitted, but also the bearings supporting the gear shafts must be of adequate size and design to sustain the radial and thrust loads that will be imposed on them. Assuming that suitable gear and pinion proportions have been selected, the next step is to compute the loads needed to determine whether or not adequate bearings can be provided. To find the loads on the bearings, first, use the formulas on the following pages to compute the tangential, axial, and separating components of the load on the tooth surfaces. Second, use the principle of moments, together with the components determined in the first step, to find the radial loads on the bearings. To illustrate the procedure, the following example will be used.

Example 25: A 16-tooth left-hand spiral pinion rotating clockwise at 1800 rpm transmits 71 horsepower to a 49-tooth mating gear. If the pressure angle is 20 degrees, the spiral angle is 35 degrees, the face width is 1.5 inches, and the diametral pitch is 5, what are the radial and thrust loads that govern the selection of bearings?

In **Fig. 3**, the locations of the bearings for the gear shafts are shown. It should be noted that distances *J*, *K*, *L*, and *M* are measured from the center line of the bearings and from the midfaces of the gears at their mean pitch diameters. In this example, it will be assumed that these distances are given and are as follows: $J = 3.5$ inches; $K = 2.5$ inches; $L = 1.5$ inches; and $M = 5.0$ inches.

Also given:

Number of pinion teeth, n	= 16	(1)
Number of gear teeth, N	= 49	(2)
Diametral pitch, P	= 5	(3)
Face width, F	= 1.5	(4)
Pressure angle, $\phi = 20°$	= 20°	(5)
Shaft angle, $\Sigma = 90°$	= 90°	(6)

Table 3. Formulas for Gleason System 20-Degree Pressure Angle, Spiral Bevel Gears—90-Degree Shaft Angle

No	Item	Formula	
		Pinion	Gear
7	Working Depth	$h_k = \frac{1.700}{P}$	Same as pinion
8	Whole Depth	$h_t = \frac{2.188}{P}$	Same as pinion
9	Pitch Diameter	$d = \frac{n}{P}$	$D = \frac{N}{P}$
10	Pitch Angle	$\gamma = \tan^{-1}\frac{n}{N}$	$\Gamma = 90° - \gamma$
11	Cone Distance	$A_O = \frac{D}{2\sin\Gamma}$	Same as pinion
12	Circular Pitch	$p = \frac{3.1416}{P}$	Same as pinion
13	Addendum	$a_p = h_k - a_G$	$a_G = \frac{0.540}{P} + \frac{0.390}{P\left(\frac{N}{n}\right)^2}$
14	Dedendum	$b_p = h_t - a_p$	$b_G = h_t - a_G$
15	Clearance	$c = h_t - h_k$	Same as pinion
16	Dedendum Angle	$\delta_P = \tan^{-1}\frac{b_p}{A_O}$	$\delta_G = \tan^{-1}\frac{b_G}{A_O}$
17	Face Angle of Blank	$\gamma_O = \gamma + \delta_G$	$\Gamma_O = \Gamma + \delta_p$

Table 3. *(Continued)* **Formulas for Gleason System 20-Degree Pressure Angle, Spiral Bevel Gears—90-Degree Shaft Angle**

No	Item	Formula	
		Pinion	Gear
18	Root Angle	$\gamma_R = \gamma - \delta_p$	$\Gamma_R = \Gamma - \delta_G$
19	Outside Diameter	$d_O = d + 2a_p \cos\gamma$	$D_O = D + 2a_G \cos\Gamma$
20	Pitch Apex to Crown	$x_O = \frac{D}{2} - a_p \sin\gamma$	$X_O = \frac{d}{2} - a_G \sin\Gamma$
21	Circular Thickness	$t = p - T$	$T = \frac{(1.5708 - K)}{P} - \frac{\tan\phi}{\cos\psi}(a_p - a_G)$
22	Backlash[a]	B = (See table on *Handbook* page **2245**)	

[a] When the gear is cut spread-blade, all the backlash is taken from the pinion thickness. When both members are cut single-side, each thickness is reduced by half of the backlash.

All linear dimensions are in inches.

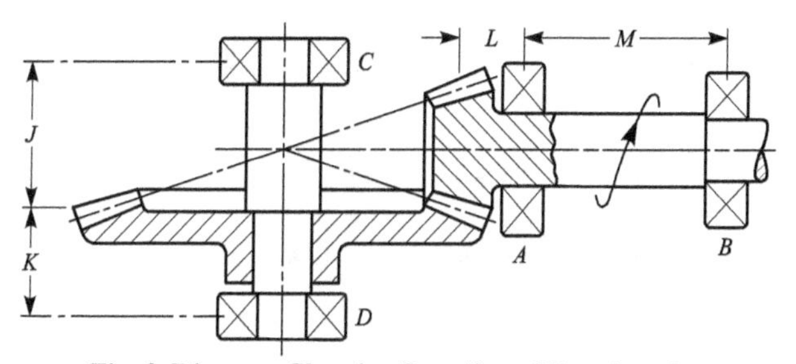

Fig. 3. Diagram Showing Location of Bearings for Bevel Gear Drive in Example 25

Other quantities that will be required in the solution of this example are the pitch diameter, pitch angle, and mean pitch diameter of both the gear and pinion. These are computed using formulas given in **Table 3** on the previous page as follows:

By using Formula 9 in **Table 3**,

Pitch dia. of pinion $d = 3.2$ inches

Pitch dia. of gear $D = 9.8$ inches

By using Formula 10 in **Table 3**,

Pitch angle of pinion $\gamma = 18° 5'$

Pitch angle of gear $\Gamma = 71° 55'$

By using the formula given below,

Mean pitch diameter of pinion

$$d_m = d - F\sin\gamma$$
$$= 3.2 - 1.5 \times 0.31040$$
$$= 2.734 \text{ inches}$$

Mean pitch diameter of gear

$$D_m = D - F\sin\Gamma$$
$$= 9.8 - 1.5 \times 0.95061$$
$$= 8.374 \text{ inches}$$

The first step in determining the bearing loads is to compute the tangential W_t, axial W_x, and separating W_s, components of the tooth load, using the formulas that follow.

$$W_t = \frac{126,050P}{nd_m} = \frac{126,050 \times 71}{1800 \times 2.734} = 1819 \text{ pounds}$$

$$W_x(\text{pinion}) = \frac{W_t}{\cos\psi}(\tan\phi\sin\gamma_d + \sin\psi\cos\gamma_d)$$
$$= \frac{1819}{0.81915}(0.36397 \times 0.31040 + 0.57358 \times 0.95061)$$
$$= 1462 \text{ pounds}$$

$$W_x\ (\text{gear}) = \frac{W_t}{\cos\psi}(\tan\phi\sin\gamma_D - \sin\psi\cos\gamma_D)$$
$$= \frac{1819}{0.81915}(0.36397 \times 0.95061 - 0.57358 \times 0.31040)$$
$$= 373 \text{ pounds}$$

$$W_s(\text{pinion}) = \frac{W_t}{\cos\psi}(\tan\phi\cos\gamma_d - \sin\psi\cos\gamma_d)$$
$$= \frac{1819}{0.81915}(0.36397 \times 0.95061 - 0.57358 \times 0.31040)$$
$$= 373 \text{ pounds}$$

$$W_s \text{ (gear)} = \frac{W_t}{\cos\psi}(\tan\phi\cos\gamma_D + \sin\psi\cos\gamma_D)$$

$$= \frac{1819}{0.81915}(0.36397 \times 0.31040 + 0.57358 \times 0.95061)$$

$$= 1462 \text{ pounds}$$

The axial thrust load on the bearings is equal to the axial component of the tooth load W_x. Since thrust loads are always taken up at only one mounting point, either bearing A or bearing B must be a bearing capable of taking a thrust of 1462 pounds, and either bearing C or bearing D must be capable of taking a thrust of 373 pounds.

The next step is to determine the magnitudes of the radial loads on the bearings A, B, C, and D. For an overhung mounted gear, or pinion, it can be shown, using the principle of moments, that the radial load on bearing A is:

$$R_A = \frac{1}{M}\sqrt{[W_t(L+M)]^2 + [W_s(L+M) - W_x r]^2} \qquad (1)$$

And the radial load on bearing B is:

$$R_B = \frac{1}{M}\sqrt{(W_t L)^2 + (W_s L - W_x r)^2} \qquad (2)$$

For a *straddle mounted gear* or pinion the radial load on bearing C is:

$$R_C = \frac{1}{J+K}\sqrt{(W_t K)^2 + (W_s K - W_x r)^2} \qquad (3)$$

And the radial load on bearing D is:

$$R_D = \frac{1}{J+K}\sqrt{(W_t J)^2 + (W_s J + W_x r)^2} \qquad (4)$$

In these formulas, r is the mean pitch radius of the gear or pinion.

These formulas will now be applied to the gear and pinion bearings in the example. An overhung mounting is used for the pinion, so **Formulas** (1) and (2) are used to determine the radial loads on the pinion bearings:

$$R_A = \frac{1}{5}\sqrt{[1819(1.5+5)]^2 + [373(1.5+5) - 1462 \times 1.367]^2}$$
$$= 2365 \text{ pounds}$$

$$R_B = \frac{1}{5}\sqrt{(1819 \times 1.5)^2 + [373 \times 1.5 - 1462 \times 1.367]^2}$$
$$= 618 \text{ pounds}$$

Because of the straddle mounting used for the gear, **Formulas** (3) and (4) are used to determine the radial loads on the gear bearings:

$$R_C = \frac{1}{3.5+2.5}\sqrt{(1819 \times 2.5)^2 + (1462 \times 2.5 - 373 \times 4.187)^2}$$
$$= 833 \text{ pounds}$$

$$R_D = \frac{1}{3.5+2.5}\sqrt{(1819 \times 3.5)^2 + (1462 \times 3.5 + 373 \times 4.187)^2}$$
$$= 1533 \text{ pounds}$$

These radial loads, and the thrust loads previously computed, are then used to select suitable bearings from manufacturers' catalogs.

It should be noted, in applying **Formulas** (1) to (4), that if both gear and pinion had overhung mountings, then **Formulas** (1) and (2) would have been used for both; if both gear and pinion had straddle mountings, then **Formulas** (3) and (4) would have been used for both. In any arrangement, the dimensions and loads for the corresponding member must be used. Also, in applying the formulas, the computed values of W_x and W_s, if they are negative, must be used in accordance with the rules applicable to negative numbers.

Gear Strength Calculations.—Methods of calculating the strength and power capacity for gears used in all types of applications are provided in American Gear Manufacturers Association (AGMA) Standards. These Standards are revised as needed by improvements in gear materials, calculation methods, and increased field experience with typical designs and application factors.

AGMA Standard 2001-B88, "Fundamental Rating Factors and Calculation Methods for Involute Spur and Helical Gear Teeth," is a revision of, and supersedes, AGMA 218.01.

The AGMA Standard presents general formulas for rating the pitting resistance and the bending strength of spur and helical involute gear teeth. It is intended to establish a common base for rating various types of gears for differing applications and to encourage the maximum practical degree of uniformity and consistency between rating practices in the gear industry. The Standard provides the basis from which more detailed AGMA Application Standards are developed and is a means for calculation of approximate ratings in the absence of such Standards. Where applicable AGMA Standards exist, they should be used in preference to this Standard. Where no application Standard exists, numerical values may be estimated for the factors used in the general equations presented in the Standard. The values of these factors may vary significantly, depending on the application, system effects, gear accuracy, manufacturing practice, and definition of what constitutes gear failure.

Information on geometry factors used in pitting resistance independent strength calculations for AGMA 908-B89, "Geometry Factors for Determining the Pitting Resistance and Bending Strength of Spur, Helical, and Herringbone Gear Teeth," is used in conjunction with AGMA 2001-B88 formulas.

PRACTICE EXERCISES FOR SECTION 19

(See *Answers to Practice Exercises for Section 19* on page **250**)

1) A spur gear of 6 diametral pitch has an outside diameter of 3.3333 inches. How many teeth has it? What is the pitch diameter? What is the tooth thickness measured along the pitch circle?

2) A gear of 6 diametral pitch has 14 teeth. Find the outside diameter, the pitch diameter, and the addendum.

3) When is the 25-degree tooth form standard preferred?

4) What dimension does a gear-tooth vernier caliper measure?

5) What are the principal 20-degree pressure angle tooth dimensions for the following diametral pitches: 4; 6; 8; 18?

6) Give the important 14½ degree pressure angle tooth dimensions for the following circular pitches: ½ inch; ¾ inch; 9⁄16 inch.

7) What two principal factors are taken into consideration in determining the power-transmitting capacity of spur gears?

8) The table on *Handbook* page **2231** shows that a No. 8 formed cutter (involute system) would be used for milling either a 12- or 13-tooth pinion, whereas a No. 7 would be used for tooth numbers from 14 to 16, inclusive. If the pitch is not changed, why is it necessary to use different cutter numbers?

9) Are hobs made in series or numbers for each pitch similar to formed cutters?

10) If the teeth of a gear have a 6⁄8 pitch, what name is applied to the tooth form?

11) A stub-tooth gear has 8⁄10 pitch. What do the figures 8 and 10 indicate?

12) What is the module of a gear?

13) Explain the use of the table of chordal thicknesses on *Handbook* page **2224**.

14) Give the dimensions of a 20-degree stub tooth of 12 pitch.

15) What are the recommended diametral pitches for fine-pitch standard gears?

16) What tooth numbers could be used in pairs of gears having the following ratios: 0.2642; 0.9615?

17) What amount of backlash is provided for general-purpose gearing, and how is the excess depth of cut to obtain it calculated?

18) What diametral pitches correspond to the following modules: 2.75; 4; 8?

19) Can bevel gears be cut by formed milling cutters?

20) Can the formed cutters used for cutting spur gears also be used for bevel gears?

21) What is the pitch angle of a bevel gear?

22) What does the term "miter" mean as applied to bevel gears?

23) What is the difference between the terms "whole depth" and "working depth" as applied to gear teeth?

24) Why do preshaved gears have a greater dedendum than gears that are finish-hobbled?

25) Are gear teeth of 8 diametral pitch larger or smaller than teeth of 4 diametral pitch, and how do these two pitches compare in regard to tooth depth and thickness?

26) Where is the pitch diameter of a bevel gear measured?

27) What is the relation between the circular pitch of a wormgear and the linear pitch of the mating worm?

28) In what respect does the helix angle of a worm differ from the helix angle of a helical or spiral gear?

29) How do the terms "pitch" and "lead," as applied to a worm, compare with the same terms as applied to screw threads?

30) Why is the outside diameter of a hob for cutting a wormgear somewhat larger than the outside diameter of the worm?

31) Why are triple, quadruple, or other multiple-threaded worms used when an efficient transmission is required?

32) In designing worm drives having multi-threaded worms, it is common practice to select a number of wormgear teeth that is not an exact multiple of the number of worm threads. Why is this done? When should this practice be avoided?

33) Explain the following terms used in connection with helical or spiral gears: "transverse diametral pitch" and "normal diametral pitch." What is the relation between these?

34) Are helical gear calculations based upon diametral pitch or circular pitch?

35) Can helical gears be cut with the formed cutters used for spur gears?

36) In spiral gearing, the tangent of the tooth or helix angle = the circumference ÷ lead. Is this circumference calculated from the outside diameter, the pitch diameter, or the root diameter?

37) What advantages are claimed for gearing of the herringbone type?

SECTION 20

SPEEDS, FEEDS, AND MACHINING POWER

Machinery's Handbook pages **1065–1141** and **1144–1183**

Metal-cutting operations such as turning and drilling may not be as productive as they could be unless the material removal rate is at or near the maximum permitted by the available power of the machine. It is not always possible to use the machine's full power owing to limitations imposed by a combination of part configuration, part material, tool material, surface finish and tolerance requirements, coolant employed, and tool life. However, even with such restrictions, it is practical to find a combination of depth of cut, feed rate, and cutting speed to achieve the best production rate for the job at hand.

The information on *Handbook* pages **1065–1141** is useful in determining how to get the most out of machining operations. The tabular data are based on actual shop experience and extensive testing in machining laboratories. A list of machining data tables is given on *Handbook* page **1078**, and these tables are referred to in the following.

Machining operations such as milling, drilling, and turning using very small tooling requires special consideration as such tools easily break, even at conservative values of cutting speed, feed, and depth of cut of conventional machining. The information on *Handbook* pages **1144–1183** explores the requirements and techniques of sucessful micromachining operations.

Most materials can be machined over a wide range of speeds; however, there is usually a narrower spread of speeds within which the most economical results are obtained. This narrower spread is determined by the economical tool life for the job at hand as, for example, when a shorter tool life is tolerable the speed can be increased. On the other hand, if tool life is too short, causing excessive down time, then speed can be reduced to lengthen tool life.

To select the best cutting conditions for machining a part the following procedure may be followed:

1) Select the maximum depth of cut consistent with the job.

2) Select the maximum feed rate that can be used consistent with such job requirements as surface finish and the rigidity of the cutting tool, workpiece, and the machine tool. Use **Table 15a** on *Handbook* page **1110** to assist in feed selection for milling. When possible, use the combined feed/speed portions of the tables to select two pairs of feed and speed data and determine the spindle speed as illustrated by **Example 1**.

3) If the combined feed/speed data are not used, select the cutting speed and determine the spindle speed (for turning also use **Table 5a** on *Handbook* page **1091**). This order of selection is based on the laws governing tool life; i.e., the life of a cutting tool is affected most by the cutting speed, then by the feed, and least by the depth of cut.

By using the same order of selection, when very heavy cuts are to be taken, the cutting speed that will utilize the maximum power available on the machine tool can be estimated by using a rearrangement of the machining power formulas on *Handbook* pages **1133–1137**. These formulas are used together with those on *Handbook* pages **1072** and **1097** which are used when taking ordinary cuts, as well as heavy cuts. Often, the available power on the machine will limit the size of the cut that can be taken. The maximum depth of cut and feed should then be used and the cutting speed adjusted to utilize the maximum available power. When the cutting speed determined in this manner is equal to or less than recommended, the maximum production and the best possible tool life will be achieved. When the estimated cutting speed is greater than recommended, the depth of cut or feed may be increased, but the cutting speed should not be increased beyond the value that will provide a reasonable tool life.

Example 1: An ASTM Class 25 (160–180 Brinell Hardness Number, BHN, which also may be expressed as HB) gray-iron casting is to be turned on a geared head lathe using a cemented carbide cutting tool. The heaviest cut will be 0.250 inch (6.35 mm) deep, taken on an 8-inch (203.2-mm) diameter of the casting; a feed rate of 0.020 in/rev (0.51 mm/rev) is selected for this cut. Calculate the spindle speed of the lathe, and estimate the power required to take this cut.

Locate the selected work material in **Table 4a** (*Handbook* page **1089**), and select the feed/speed pairs that correspond to the chosen cutter material. For an uncoated carbide tool, the given feed/speed pairs are: optimum 28/240, and average 13/365.

Factors to correct for feed and depth of cut are found in **Table 5a** (page **1091**). First, determine the ratios of $^{\text{chosen feed}}/_{\text{optimum feed}} = {}^{20}/_{28} =$ 0.71 and $V_{avg}/V_{opt} = {}^{365}/_{240} = 1.52$, then, by estimation or interpolation, determine F_f and F_d, and calculate V and N as follows:

$F_f = 1.22; F_d = 0.86$

$V = V_{opt} \times F_f \times F_d = 240 \times 1.22 \times 0.86 = 252$ ft/min

$$N = \frac{12V}{\pi D} = \frac{12 \times 252}{\pi \times 8} = 120 \text{ rpm}$$

See tables from *Handbook* pages **1133** on to estimate the power requirements using: $K_p = 0.52$ (**Table 1a**), $C = 0.90$ (**Table 2**), $Q = 12Vfd$ (**Table 5**), $W = 1.30$ (**Table 3**), and $E = 0.80$ (**Table 4**).

$Q = 12Vfd = 12 \times 252 \times 0.020 \times 0.250 = 15.12\ \text{in}^3/\text{min}$

$$P_m = \frac{K_p CQW}{E} = \frac{0.52 \times 0.90 \times 15.12 \times 1.30}{0.80} = 11.5 \text{ hp}$$

The equivalent results, expressed in the metric system, can be obtained by converting the cutting speed V, the metal removal rate Q, and the power at the motor P_m into metric units using factors found starting on page **2845** of the *Handbook*, as illustrated in the following.

$V = 252$ ft/min $= 252 \times 0.3 = 76$ m/min

$Q = 15.12\ \text{in}^3/\text{min} = 15.12 \times 16.4 \div 60 = 4.13\ \text{cm}^3/\text{s}$

$P_m = 11.5$ hp $= 11.5 \times 0.745 = 8.6$ kw

Alternatively, if metric units are used throughout the problem, F_f and F_d are determined in the same manner as above. However, if V is in meters per minute, and D and d are in millimeters, then $N = 1000V/\pi D$, and $Q = Vfd/60$.

Example 2: If the lathe in **Example 1** has only a 10-hp motor, estimate the cutting speed and spindle speed that will utilize the maximum available power. Use inch units only.

$$Q_{max} = \frac{P_m E}{K_p CW} = \frac{10 \times 0.80}{0.52 \times 0.90 \times 1.30} \qquad \left(P_m = \frac{K_p CQW}{E}\right)$$

$$= 13.15(\text{in}^3/\text{min})$$

$$V = \frac{Q_{max}}{12fd} = \frac{13.15}{12 \times 0.020 \times 0.250} \qquad (Q = 12Vfd)$$

$$= 219 \text{ fpm}$$

$$N = \frac{12V}{\pi D} = \frac{12 \times 219}{\pi \times 8} = 105 \text{ rpm}$$

Example 3: A slab milling operation is to be performed on 120–140 BHN AISI 1020 steel using a 3-inch diameter high-speed steel plain milling cutter having 8 teeth. The width of this cut is 2 inches; the depth is 0.250 inch, and the feed rate is 0.004 in/tooth. Estimate the power at the motor required to take this cut (except for **Table 11**, refer to tables starting on *Handbook* page **1135**).

$V = 110$ fpm (**Table 11**, page **1101**) $Q = f_m wd$ (**Table 5**)
$K_p = 0.69$ (**Table 1b**) $W = 1.10$ (**Table 3**)
$C = 1.25$ (**Table 2**) $E = 0.80$ (**Table 4**)

$$N = \frac{12V}{\pi D} = \frac{12 \times 110}{\pi \times 3} = 140 \text{ rpm}$$

$$f_m = f_t n_t N = 0.004 \times 8 \times 140 = 4.5 \text{ in/min}$$

$$P_m = \frac{K_p CQW}{E} = \frac{0.69 \times 1.25 \times 2.25 \times 1.10}{0.80} = 2.67 \text{ hp}$$

Example 4: A 16-inch-diameter cemented carbide face milling cutter having 18 teeth is to be used to take a 14-inch-wide and 0.125-inch-deep cut on an H12 tool steel die block having a hardness of 250–275 BHN. The feed used will be 0.008 in/tooth, and the milling machine has a 20-hp motor. Using the tables starting on *Handbook* page **1133**, estimate the cutting speed and the spindle speed to be used that will utilize the maximum horsepower available on the machine.

$K_p = 0.98$ fpm (**Table 1a**) $W = 1.25$ (**Table 3**)
$C = 1.08$ (**Table 2**) $E = 0.80$ (**Table 4**)
$Q = f_m wd$ (**Table 5**)

$$Q_{max} = \frac{P_m E}{K_p C W} = \frac{20 \times 0.80}{0.98 \times 1.08 \times 1.25} \qquad \left(P_m = \frac{K_p C Q W}{E}\right)$$

$$= 12.1 \ \text{(in}^3\text{/min)}$$

$$f_m = \frac{Q_{max}}{wd} = \frac{12}{14 \times 0.125} \qquad (Q = f_m w d)$$

$$= 6.9 \ \text{in/min; use 7 in/min}$$

$$N = \frac{f_m}{f_t n_t} = \frac{7}{0.008 \times 18} \qquad (f_m = f_t n_t N)$$

$$= 48.6 \ \text{rpm; use 50 rpm}$$

$$V = \frac{\pi D N}{12} = \frac{\pi \times 16 \times 50}{12} = 209 \ \text{fpm}$$

Formulas for estimating the thrust, torque, and power for drilling are given on *Handbook* page **1139**. Thrust is the force required to push or feed the drill when drilling. This force can be very large. It is sometimes helpful to know the magnitude of this force and the torque exerted by the drill when designing drill jigs or work-holding fixtures; it is essential to have this information as well as the power required to drill when designing machine tools on which drilling operations are to be performed. In the ordinary shop, it is often helpful to be able to estimate the power required to drill larger holes in order to determine if the operation is within the capacity of the machine to be used.

Example 5: Estimate the thrust, torque, and power at the motor required to drill a ³⁄₄-inch-diameter hole in a part made from AISI 1117 steel, using a conventional twist drill and a feed rate of 0.008 in/rev. (Except for **Table 17**, see *Handbook* pages **1136** on for tables.)

K_d = 12,000 (**Table 6**) B = 1.355 (**Table 7**)
F_f = 0.021 (**Table 8**) J = 0.030 (**Table 7**)
F_T = 0.794 (**Table 9**) E = 0.80 (**Table 4**)
F_M = 0.596 (**Table 9**) W = 1.30 (**Table 3**)
A = 1.085 (**Table 7**) V = 101 fpm (**Table 17**, page **1117**)

$$T = 2K_dF_fF_TBW + K_dd^2JW$$
$$= 2 \times 12{,}000 \times 0.021 \times 0.794 \times 1.355 \times 1.30 + 12{,}000 \times 0.75^2 \times 0.030 \times 1.30$$
$$= 968 \text{ lb}$$

$$M = K_dF_fF_MAW$$
$$= 12{,}000 \times 0.021 \times 0.596 \times 1.085 \times 1.30$$
$$= 212 \text{ in-lb}$$

$$N = \frac{12V}{\pi D} = \frac{12 \times 101}{\pi \times 0.750} = 514 \text{ rpm}$$

$$P_c = \frac{MN}{63{,}025} = \frac{212 \times 514}{63{,}025} = 1.73 \text{ hp}$$

$$P_m = \frac{P_c}{E} = \frac{1.73}{0.80} = 2.16 \text{ hp}$$

PRACTICE EXERCISES FOR SECTION 20

(See *Answers to Practice Exercises for Section 20* on page **252**)

1) Calculate the spindle speeds for turning $\frac{1}{2}$-inch and 4-inch bars made from the following steels, using a high-speed steel cutting tool and the cutting conditions given as follows:

Steel Designation	Feed, in/rev	Depth of Cut, inch
AISI 1108, Cold Drawn	0.012	0.062
12L13, 150–200 BHN	0.008	0.250
1040, Hot Rolled	0.015	0.100
1040, 375–425 BHN	0.015	0.100
41L40, 200–250 BHN	0.015	0.100
4140, Hot Rolled	0.015	0.100
O2, Tool Steel	0.012	0.125
M2, Tool Steel	0.010	0.200

2) Calculate the spindle speeds for turning 6-inch-diameter sections of the following materials, using a cemented carbide cutting tool and the cutting conditions given in the table on the next page.

Material	Feed, in/rev	Depth of Cut, inch
AISI 1330, 200 BHN	0.030	0.150
201 Stainless Steel, Cold Drawn	0.012	0.100
ASTM Class 50 Gray Cast Iron	0.016	0.125
6A1-4V Titanium Alloy	0.018	0.188
Waspaloy	0.020	0.062

3) A 200 BHN AISI 1030 forged steel shaft is being turned at a constant spindle speed of 400 rpm, using a cemented carbide cutting tool. The as-forged diameters of the shaft are 1½, 3, and 4 inches. Calculate the cutting speeds (fpm) at these diameters, and check to see if they are within the recommended cutting speed.

4) A 75-mm diameter bar of cold-drawn wrought aluminum is to be turned with a high-speed steel cutting tool, using a cutting speed of 180 m/mm. Calculate the spindle speed that should be used.

5) Calculate the spindle speed required to mill a 745 nickel-silver part using a ½-inch end milling cutter.

6) An AISI 4118 part having a hardness of 200 BHN is to be machined on a milling machine. Calculate the spindle speeds for each of the operations below and the milling-machine table feed rates for Operations a) and b).

a) Face mill top surface, using an 8-inch diameter cemented-carbide face milling cutter having 2 teeth. (Use $f_t = 0.008$ in/tooth.)

b) Mill ¼-inch-deep slot, using a ¾-inch-diameter two-fluted high-speed steel end milling cutter.

c) Drill a 23/64-inch hole.

d) Ream the hole ⅜ inch, using HSS reamer.

7) A 3-inch-diameter high-speed steel end milling cutter having 12 teeth is used to mill a piece of D2 high carbon, high chromium cold-work tool steel having a hardness of 220 BHN. The spindle speed used is 75 rpm, and the milling-machine table feed rate is 10 in/mm. Check the cutting conditions with respect to the recommended values, and make recommendations for improvements, if possible.

8) A 100–150 BHN low-carbon steel casting is to be machined with a 12-inch-diameter cemented carbide face milling cutter having 14 teeth, using a spindle speed of 60 rpm and a table-feed rate of 5 in/mm. Check these cutting conditions and recommend improvements, if possible.

9) Estimate the cutting speed and the power at the cutter and at the motor required to turn 210 BHN AISI 1040 steel in a geared head lathe, using an uncoated carbide tool, a depth of cut of 0.125 inches, a feed of 0.015 in/rev, and efficiency E of 0.80.

10) A 165 BHN A286 high-temperature alloy, or superalloy, is to be turned on a 3-hp geared head lathe using a cemented carbide cutting tool. The depth of cut selected is 0.100 inch, and the feed is 0.020 in/rev. Estimate the cutting speed that will utilize the maximum power available on the lathe.

11) An AISI 8642 steel having a hardness of 210 BHN is to be milled with a 6-inch-diameter cemented carbide face milling cutter having 8 teeth on a 10-hp milling machine. The depth of cut is to be 0.200 inch, the width is 4 inches, and the feed is to be 0.010 in/tooth. Estimate the cutting speed that will utilize the maximum power available on the machine.

12) Estimate the thrust, torque, and power at the motor required to drill 200 BHN steel using the following drill sizes, feeds, and spindle speeds.

Drill Size	Feed	Spindle Speed
$\frac{1}{4}$ in.	0.0005 in/rev	1500 rpm
$\frac{1}{2}$ in.	0.002 in/rev	750 rpm
1 in.	0.008 in/rev	375 rpm
19 mm	0.15 mm/rev	500 rpm

13) Estimate the thrust, torque, and power at the motor for the 1-inch drill in Exercise 12 if the drill is ground to have a split point.

14) Describe the general characteristics of high-speed steels that make them suitable for use as cutting-tool materials.

15) What guidelines should be followed in selecting a grade of cemented carbide?

16) How does the cutting speed, feed, and depth of cut influence tool life?

17) List the steps for selecting the cutting conditions in their correct order and explain why.

18) What are the advantages of coated carbides, and how should they be used?

19) Name the factors that must be considered when selecting a cutting speed for tapping.

20) Why is it important to calculate the table-feed rate for milling?

21) Name the factors that affect the basic feed rate for milling.

22) When should the power required to take a cut be estimated? Why?

23) Name the factors that affect the power constant, K_p. This constant is unaffected by what?

24) Why is it necessary to have a separate method for estimating the drilling thrust, torque, and power?

25) Why are traditional speeds and feeds generally inappropriate for micromachining?

26) What is pecking or peck drilling and why is it used?

SECTION 21

CNC (COMPUTER NUMERICAL CONTROL) PROGRAMMING

Machinery's Handbook pages **1338–1387**

Numerical control (NC) was defined by the Electronic Industries Association as "a system in which actions are controlled by the direct insertion of numerical data at some point. The system must automatically interpret at least some portion of these data." From this concept evolved modern manufacturing methods using computer numerical control programming, better known as *CNC programming*.

CNC manufacturing can be defined as using coded instructions (expressed with letters in the English alphabet, numerals, and symbols) submitted to and processed by a programmable machine control system, which then operates a machine's tools to modify or create a workpiece. All instructions must be written in a logical order, in a specified format called the *program structure*, taking into account the CNC machine's setup. Individual program instructions include all required machine activities: tool selection, motion control, spindle speed, cutting feed rates, coolant flow, various adjustments, and so on. The part program can be stored and used repeatedly to produce the same part with minimal human interaction.

The CNC programmer is responsible for gathering and integrating all pertinent information involved in developing safe, efficient, error-free, production-ready part programs for CNC machines installed in the shop. The programmer needs to know the machines and tools, visualize all tool motions, and, understanding the materials and part parameters, recognize any restricting factors. Even with the help of today's automated software, knowledge of programming is essential, as is effective communication among collaborators on any parts project.

Information in the *Handbook*, starting on page **1338**, gives an overview of the most important CNC concepts and operations. Additionally, material on numerical control programming from previous editions of the *Handbook*, starting on *Machinery's Handbook 32 Digital Edition* page **3430**, is arranged by subject matter for ease of reference. (Note that as with all *Digital Edition*–only material, this section has not been updated.) A related older discussion—of the Automatic Programmed Tool (APT) language, its use in part programming, and examples of computational and geometric programs—also can be found starting on *Machinery's Handbook 32 Digital Edition* page **3455**, though APT programming of this type is no longer required with today's CNC machines.

CNC Coordinate Geometry.—Key to working with CNC principles is understanding the *geometry* of machine tools. This encompasses the relationship between the machine data, part data, and tool data, which includes setup (see *Handbook* page **1339**).

As with all CAD/CAM systems (see *CAD/CAM* on *Handbook* page **1377**), CNC is based on the same principles of a system of coordinates that defines the location of a point in two-dimensional (2D) or three-dimensional space (3D). A system of coordinates is founded on the concept of two perpendicular lines (named axes "X" and "Y") intersecting at a point called the *origin*, where both coordinates have a value of zero (X0, Y0). Both lines are divided into equal units of measurement as shown in **Fig. 1**.

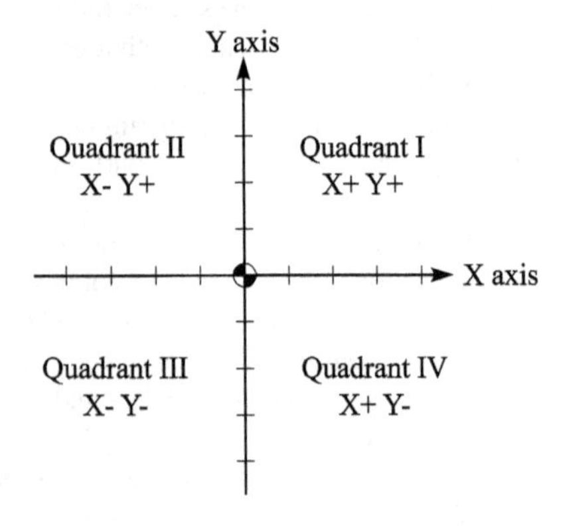

Fig. 1. Rectangular Coordinate System

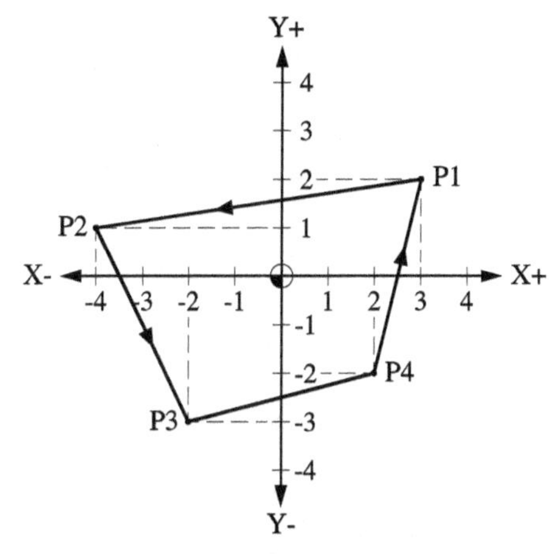

Fig. 2. Absolute Coordinates and Incremental Linear Motion

A point can be defined in a plane (any two axes = 2D) or in space (three axes = 3D). For CNC work, the *rectangular coordinate system*—also known as the *Cartesian coordinate system*—is the most commonly used. It is based on three standard axes: X,Y,Z. *Points* are defined as locations with a distance from the origin defined by projecting a line at 90 degrees to each axis, forming a visual rectangle. For example, in **Fig. 2**, P1 is defined as X3.0, Y2.0, P2 is defined as X-4.0, Y1.0, P3 as X-2.0,Y-3.0, and P4 as X2.0, Y-2.0.

Once the point locations are established, the programmer can use these points to specify the center of individual holes to be machined or as endpoints for a continuous contour. These points also can be connected by providing specific motion instructions, as a toolpath between given points.

Point-to-Point Programming.—Point-to-point programming is covered beginning on *Handbook* page **1354**. As an example of the use of CNC for point-to-point part programs, consider the rectangular plate shown in **Fig. 3**, in which it is required to machine eight holes as shown.

Dimensions for the positions of the holes are provided based on their distances from the *X* and *Y* axes—the origin, in this case, located at the center of the part. This positioning information is entered into the CNC part program, along with instructions for the tooling to be loaded into the spindle for the work to be performed. The hole coordinate location given on the design drawing, as shown in the table following **Fig. 3** below, is entered in the part programming manuscript, together with coded details, such as spindle speed and feed rates.

All of this information and coded instructions, according to the special codes discussed on pages **1343–1348** of the *Handbook*, direct precise machining of the holes when the CNC program is executed.

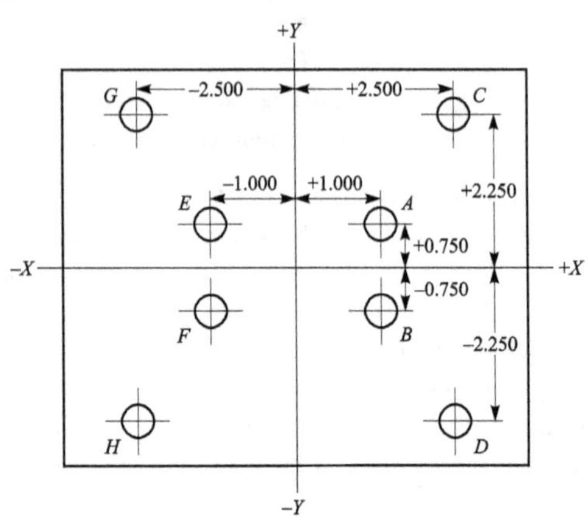

Fig. 3. Alternative Methods of Dimensioning for the Positions of Eight Holes to Be Machined in a Rectangular Plate

Hole Coordinate Location

	Dimensions on Axes			Dimensions on Axes	
Point	*X*	*Y*	Point	*X*	*Y*
A	+ 1.000	+ 0.750	*E*	– 1.000	+ 0.750
B	+ 1.000	– 0.750	*F*	– 1.000	– 0.750
C	+ 2.500	+ 2.250	*G*	– 2.500	+ 2.250
D	+ 2.500	– 2.250	*H*	– 2.500	– 2.250

Absolute and Incremental Programming.—A motion dimension along the axes *X*, *Y*, and *Z* for milling and axes *X* and *Z* for turning (as well as any parallel or rotary axis) can be programmed either as an *absolute location* measured from part zero (origin) or as an *incremental distance* (*U* and *W* for turning) with direction measured from the current tool position. The absolute method of programming is selected by the G90 command, and the incremental method is selected by the G91 command as described on *Handbook* page **1347**. (Each command cancels the other.)

The benefit of absolute programming is that one change of a point location requires only one change in the program; the same change in incremental programming requires two changes. The main benefit of incremental programming is that it can be used for toolpath repetition, typically in subprograms or macros.

Continuous-Path Programming.—Surfaces along or at angles to the axes and curved surfaces are produced by *continuous-path* or *contouring* programs. (See the section *Calculations for Contouring* on page **1359** of the *Handbook*.) These programs coordinate two or more machine motions simultaneously, so that the movement of the workpiece relative to the cutting tool generates the required shape.

Angular shapes are generated by *straight line* or *linear interpolation* programs that coordinate movements of two slides to produce the required angular path. Circular arcs can be generated by means of a *circular interpolation* program that controls the slide movements automatically to produce a curved path.

Linear Interpolation.—In *linear interpolation* mode (programmed in G01 mode with a cutting feed rate in effect), the tool moves along a linear path, directly from one point to the next, traversing the shortest distance, as described on *Handbook* page **1350**. (This is a departure from *rapid motion* mode, programmed in G00 mode, which instead selects the path for the fastest motion.) The word "interpolation" defines a mathematical process that translates into a synchronized feed rate for all programmed axes; this is automatically calculated and applied by the control system.

Circular Interpolation.—*Circular interpolation* applies to a cutting toolpath moving along an arc or a circle. The cutting motion along an arc is two-dimensional and takes place in a single plane. The plane in which the arc is positioned must be determined initially.

(No arc or circle exists as a 3D entity; it becomes a helix.) Depending on the required outcome, the desired plane can be defined using one of three G-codes (G17, G18, or G19), as described on *Handbook* page **1350**. CNC mills usually default to the *X-Y* axes (G17) upon startup.

Programming an arc requires a direction of motion in the plane (G02 = clockwise; G03 = counterclockwise), a start point (typically, the current block), an end point (typically, the target point at the end of the motion), and the arc radius. The arc radius can be specified directly (for example, R1.5), or it can be defined by special vectors, using I-J-K addresses (see **Fig. 4** and the definitions given below).

The definition of vectors I, J, and K is the same for all three axes, where X and I, Y and J, and Z and K are related (paired): the vector I-J-K is the incremental distance and direction from the start point of the arc to the arc center, measured along axis X-Y-Z respectively.

A typical milling application in the plane XY is shown in **Fig. 4**. Current plane selection determines the arc vectors. Some controls use I-J-K as an absolute location. The start point of the arc is defined as the endpoint of the previous motion.

When specifying a direct radius R, there are some limitations. Radius R (positive) can only be programmed for arcs up to and including a 180-degree arc sweep angle. If the sweep angle exceeds 180 degrees but is less than 360 degrees, R– (negative) must be used. For a full circle (360 degrees arc sweep angle), the address R cannot be used at all. Instead, a two-vector combination must be used; there is no need to break arc motion at the quadrants.

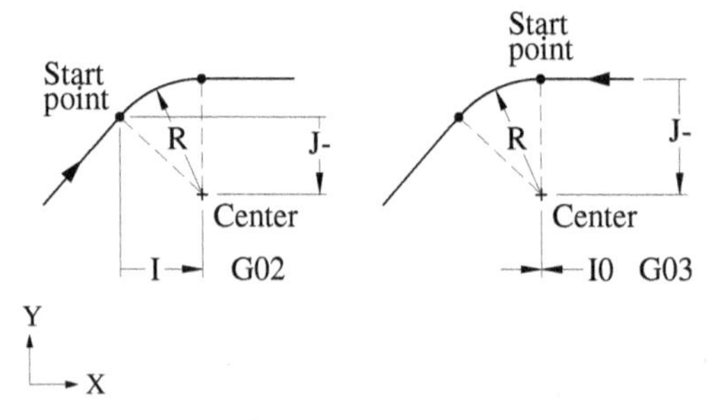

Fig. 4. Arc Vectors

PRACTICE EXERCISES FOR SECTION 21

(See *Answers to Practice Exercises for Section 21* on page **254**)

1) List the different methods of program development.

2) What type of actions are controlled in CNC programs?

3) What is the address (code beginning) used to preset or to prepare the control system for a certain desired condition, or to a certain mode or a state of operation?

4) Which of the following applications of CNC is most common? (a) grinding, (b) turning, (c) broaching, or (d) milling and turning?

5) What is the default (at machine startup) feed rate on CNC lathes measured in? (a) cutting speed, (b) inches per revolution in/rev or mm/rev, (c) inches per minute in/min or mm/min, or (d) constant surface speed?

6) What other program word is necessary when programming G01 linear interpolation? (a) S, (b) F, (c) T, or (d) H?

7) When programming an arc, which letters identify the arc center location? (a) X, Y, and Z, (b) A, B, and C, (c) I, J, and K, or (d) Q and P?

8) The letter address H is used to indicate a tool length offset register number. Which preparatory function is it used in conjunction with? (a) G54, (b) G43, (c) G42, or (d) G41?

9) When programming an arc, an additional method exists that does not use I, J, and K. Which program word is used? (a) A, (b) B, (c) C, or (d) R?

10) What is the program word "N" that identifies a program block called?

11) The advantage of using G96 Constant Cutting Speed in turning is that as the diameter changes (position of the tool changes in relation to the centerline), the r/min increases or decreases to accomplish the programmed cutting speed. True or false?

12) What is the difference between cutter radius offset and cutter compensation?

13) Which G-Code is used to cancel Cutter Diameter Compensation? (a) G40, (b) G41, (c) G42, or (d) G43?

14) Fixed cycles, sometimes called canned cycles, are a series of preset sequences of toolpath motions, allowing consecutive motions to be performed with a single block of program. What do the codes G98 and G99 represent?

15) In the first column below various fixed or canned cycles are shown. In the second column are some preparatory codes. Match the functions with the codes.

a. Spot-drilling cycle	1. G80
b. Peck-drilling cycle (deep hole)	2. G81
c. Boring cycle, spindle rotating on withdrawal at feed rate	3. G85
d. Drilling cycle	4. G84
e. Right-hand threading cycle	5. G82
f. Fixed cycle cancellation	6. G83

16) A parametric subroutine is used exclusively for describing the path around the outside of a part. True or false?

17) What is a G word?

18) Explain the rule that describes the orientation and directions of the motions of slides and spindles on a machine tool.

19) With computer-aided manufacturing (CAM), the design model from CAD is used to identify specific cutting tool data and create cutting toolpaths for CNC machining. True or false?

20) The simulation and verification of the final output of CNC code requires input of accurate setup data to perform effectively. Name some of the common criteria required.

SECTION 22

THE METRIC SYSTEM

Machinery's Handbook contains a considerable amount of metric material in terms of texts, tables, and formulas. This material is included because much of the world now uses the metric system, also known as the Système International (SI), and the movement in that direction continues in all countries that intend to compete in the international marketplace, including the United States.

An explanation of the SI metric system is found on *Handbook* pages **151–154** and **2840–2843**. A brief history is given of the development of this system, and a description is provided for each of its seven basic units. Factors and prefixes for forming decimal multiples and submultiples of the SI units also are shown. Another table lists SI units with complex names and symbols for them.

Tables of SI units and conversion factors appear on pages **2845–2883**. Factors are provided for converting English units to metric units, or vice versa, and cover units of length, area, volume (including capacity), velocity, acceleration, flow, mass, density, force, force per unit length, bending moment or torque, moment of inertia, section modulus, momentum, pressure, stress, energy, work, power, and viscosity. By using the factors in these tables, it is a simple matter of multiplication to convert from one system of units to the other. Where the conversion factors are exact, they are given to only 3 or 4 significant figures, but where they are not exact they are given to 7 significant figures to permit the maximum degree of accuracy to be obtained that is ordinarily required in the metalworking field.

To avoid the need to use some of the conversion factors, various conversion tables are given starting on page **2845**. The tables for length conversion on pages **2845–2858** will probably be the most frequently used. Two different types of tables are shown. The two tables on page **2849** facilitate converting lengths up to 100 inches into millimeters, in steps of one ten-thousandth of an inch; and up to 1000 millimeters to inches, in steps of a thousandth of a millimeter.

The table starting on page **2850** enables converting fractions and mixed number lengths up to 41 inches into millimeters, in steps of one sixty-fourth of an inch.

To make possible such a wide range in a compact table, the reader often must take two or more numbers from the table and add them together, as explained in the accompanying text. The tables starting on pages **2852** and **2854** have a much more limited range of conversion for inches to millimeters and millimeters to inches. However, these table have the advantage of being direct-reading; that is, only a single value is taken from the table, and no addition is required.

For those who are engaged in design work where it is necessary to do computations in the fields of mechanics and strength of materials, a considerable amount of guidance will be found for the use of metric units. Thus, beginning on *Handbook* page **151**, the use of the metric SI system in mechanics calculations is explained in detail. In succeeding pages, **boldface type** is used to highlight references to metric units in the combined Mechanics and Strength of Materials section. Metric formulas are provided also, to parallel the formulas for English units.

As another example, on page **212**, it is explained in boldface type that SI metric units can be applied in the calculations in place of the English units of measurement without changes to the formulas for simple stresses.

The reader also should be aware that certain tables in the *Handbook*, such as that on page **80**, which gives values for segments of circles for a radius = 1, can be used for either English or metric units, as indicated directly under the table heading. There are other instances, however, where separate tables are needed, such as are shown for the conversion of cutting speed in feet per minute into revolutions per minute on pages **1074–1075**, and cutting speed in meters per minute into revolutions per minute on pages **1076–1077**.

The metric material in the *Handbook* will provide considerable useful data and assistance to engineers and technicians who are required to use metric units of measurements. It is strongly suggested that all readers, whether or not they are using metric units at the present time, become familiar with the SI System by reading the explanatory material in the *Handbook* and by studying the SI units and the ways of converting English units to them.

SI Base Units and Definitions.—*Meter (m):* The base unit of length, equal to the path length traveled by light in a vacuum during a time interval of 3.3356410×10^{-9} s (derived from the precise measurement of light's speed, c = 299,792,458 m/s). This value was adopted in 1983, based on the newly defined value for the speed of light, made possible by the technology developed at the National Institute of Standards and Technology (NIST).

Kilogram (kg): The base unit of mass, related to the most precise measurement to date of the Planck constant h, which is the energy carried by a photon to its frequency, and measured as $6.6260702 \times 10^{-34}$ kg m^2/s^2. It is with respect to this invariant of nature that the precise standard for the kilogram is given.

Second (s): The duration of 9.1926318×10^9 periods of the radiation, corresponding to the transition between the two hyperfine levels of the ground state of the cesium-133 atom, is the element used for the atomic clock due to its stability.

Ampere (A): The base unit of electric current, measured as current flow per unit of time. One ampere (*amp*) is equal to one coulomb per second. One elementary electric charge (e) is equal to 1.60218×10^{-19} coulomb (that is, 1 C = 6.241×10^{18} e).

Candela (cd): The base unit for photometry, which is the science of measuring light perceived by the human eye. The candela (cd) is the luminous intensity, in a given direction, of a source that emits monochromatic radiation of frequency 540×10^{12} hertz. The human eye is most sensitive to this frequency (greenish-yellow light). One cd has a radiant intensity of 1/683 watt per steradian.

Kelvin (K): The base unit of thermodynamic temperature, defined by taking the fixed numerical value of the Boltzmann constant k to be 1.380649×10^{-23} $J{\cdot}K^{-1}$.

Mole (mol): The base unit for the amount of a chemical substance consisting of exactly $6.02214076 \times 10^{23}$ constitutive particles. This

Table 1. SI Base Units

Quantity	Name	Symbol
Length	meter	m
Mass	kilogram	kg
Time	second	s
Electric Current	ampere	A
Luminous Intensity	candela	cd
Thermodynamic Temperature	kelvin	K
Amount of Substance	mole	mol

number is the Avogadro constant $N_A = 6.02214076 \times 10^{23}$ mol^{-1}. (Previously, 1 mole was linked to the number of atoms in 12.1 g of the carbon-12 isotope.)

SI Derived Units.—From the base units, come the SI derived units, as shown in **Table 2**. See also the following tables, *Units of Measure and Conversion Factors* on page **268** of this *Guide*, and the *MEASURING UNITS* section of the *Handbook*, starting on page **2835**.

Table 2. SI Derived Units

Quantity	Name	Symbol
Area	square meter	m^2
Volume	cubic meter	m^3
Speed, Velocity	meter per second	m/s
Acceleration	meter per second squared	m/s^2
Wave Number	reciprocal meter	m^{-1}
Mass Density	kilogram per cubic meter	kg/m^3
Specific Volume	cubic meter per kilogram	m^3/kg
Current Density	ampere per square meter	A/m^3
Magnetic Field Strength	ampere per meter	A/m
Amount-of-Substance Concentration	mole per cubic meter	mol/m^3
Luminance	candle per square meter	cd/m^2
Mass Fraction	kilogram per kilogram	kg/kg

Table 3. SI Derived Units with Special Names and Symbols

Quantity	Name and Derived Units	Symbol
Plane Angle	radian, m/m	rad
Solid Angle	steradian, m^2/m^2	sr
Frequency	hertz, s^{-1}	Hz
Force	newton, $kg{\cdot}m{\cdot}s^{-2}$	N
Pressure, Stress	pascal, $kg{\cdot}m^{-1}{\cdot}s^{-2}$	Pa
Energy, Work, Heat	joule, $kg{\cdot}m^2{\cdot}s^{-2}$	J
Power, Radiant Flux	watt, $kg{\cdot}m^2{\cdot}s^{-3}$	W
Quantity of Electric Charge	coulomb, A·s	C
Electric Potential Difference	volt, $kg{\cdot}m^2{\cdot}s^{-3}{\cdot}A^{-1}$	V
Capacitance	farad, $s^4{\cdot}A^2{\cdot}m^{-2}{\cdot}kg^{-1}$	F
Electric Resistance	ohm, $kg{\cdot}m^2{\cdot}s^{-3}{\cdot}A^{-2}$	Ω
Electric Conductance	siemens, $kg^{-1}{\cdot}m^{-2}{\cdot}s^3{\cdot}A^2$	S
Magnetic Flux	weber, $kg{\cdot}m^2{\cdot}s^{-2}{\cdot}A^{-1}$	Wb
Magnetic Flux Density	tesla, $kg{\cdot}s^{-2}{\cdot}A^{-1}$	T
Inductance	henry, $kg{\cdot}m^2{\cdot}s^{-2}{\cdot}A^{-2}$	H
Celsius Temperature	degree Celsius, K − 273.15	°C
Luminous Flux	lumen, cd·sr	lm
Illuminance	lux, $cd{\cdot}sr{\cdot}m^{-2}$	lx
Activity of Radionuclide	becquerel, s^{-1}	Bq
Absorbed Dose	gray, $J{\cdot}kg^{-1}$	Gy
Dose Equivalent	sievert, $m^2{\cdot}s^{-2}$	Sv
Catalytic Activity	katal, $mol{\cdot}s^{-1}$	kat

Table 4. SI Derived Units that Include SI Derived Units with Special Names

Quantity	Name	Symbol
Dynamic Viscosity	pascal-second	Pa·s
Moment of Force	newton-meter	N·m
Angular Velocity	radian per second	rad/s
Angular Acceleration	radian per second squared	rad/s^2
Heat Flux Density, Irradiance	watt per square meter	W/m^2
Heat Capacity, Entropy	joule per kelvin	J/K
Specific Heat Capacity, Specific Entropy	joule per kilogram-kelvin	J/kg·K
Specific Energy	joule per kilogram	J/kg
Energy Density	joule per cubic meter	J/m^3
Thermal Conductivity	watt per meter-kelvin	W/m·K
Electric Field Strength	volt per meter	V/m
Electric Charge Density	coulomb per cubic meter	C/m^3
Electric Flux Density	coulomb per square meter	C/m^2
Permittivity	farad per meter	F/m
Permeability	henry per meter	H/m
Molar Energy	joule per mole	J/mol
Molar Entropy, Molar Heat Capacity	joule per mole-kelvin	J/mol·K
Exposure (X-rays and γ rays)	coulomb per kilogram	C/kg
Absorbed Dose Rate	gray per second	Gy/s
Radian Intensity	watt per steradian	W/sr
Radiance	watt per square meter-steradian	$W/m^2 \cdot sr$

Table 5. Units Outside SI, Accepted for Use with SI

Name	Symbol	Value in SI Units
minute	min	1 min = 60 s
hour	h	1 h = 60 min = 3600 s
day	d	1 d = 24 h = 1440 min = 86400 s
liter	L	$1 L = 1 dm^3 = 10^{-3} m$
metric ton	t	$1 t = 10^3 kg = 2205 lb$
bel	B	1 B = 10 dB
degree (angle)	°	$1° = \pi/180$ rad
minute (angle)	′	$1' = (1/60)° = (\pi/10800)$ rad
second (angle)	″	$1'' = (1/60)' = (\pi/6480000)$ rad
electron volt	eV	$1 eV = 1.60218 \times 10^{-19} J$
unified atomic mass unit	Da or u	$1 u = 1.66054 \times 10^{-27} kg$
astronomical unit	au	$1 au = 1.49598 \times 10^{11} m$
nautical mile	nmi	1 nmi = 1852 m
knot	kn	$1 kn = 1 nmi \cdot h^{-1} = 0.514444 m \cdot s^{-1}$
are	a	$1 a = 100 m^2$
hectare	ha	$1 ha = 100 a = 10^4 m^2$
bar	bar	$1 bar = 10^2 kPa = 10^5 Pa$
ångström	Å	$1 Å = 0.1 nm = 10^{-10} m$
curie	Ci	$1 Ci = 3.7 \times 10^{10} Bq$
roentgen	R	$1 R = 2.58 \times 10^{-4} C \cdot kg^{-1}$
rad	rad	$1 rad = 10^{-2} Gy$
rem	rem	$1 rem = 10^2 Sv$

Table 6. SI Prefixes

Factor	Name	Symbol	Factor	Name	Symbol
10^{1}	deca	da	10^{-1}	deci	d
10^{2}	hecto	h	10^{-2}	centi	c
10^{3}	kilo	k	10^{-3}	milli	m
10^{6}	mega	M	10^{-6}	micro	μ
10^{9}	giga	G	10^{-9}	nano	n
10^{12}	tera	T	10^{-12}	pico	p
10^{15}	peta	P	10^{-15}	femto	f
10^{18}	exa	E	10^{-18}	atto	a

SECTION 23

GENERAL REVIEW QUESTIONS

(See *Answers to General Review Questions* on page **256**)

1) If a regular polygon of 20 sides is to have an area of 100 square inches what formula may be used to calculate the length of one side of the polygon?

2) What does the number of a Jarno taper indicate?

3) What is the general rule for determining the direction in which to apply tolerances?

4) Why is 1 horsepower equivalent to 33,000 foot-pounds of work per minute? Why not 30,000 or some other number?

5) What is the chief element in the composition of babbitt metals?

6) If the pitch of a stub-tooth gear is $^{8}/_{10}$, what is the tooth depth?

7) What does the figure 8 mean if the pitch of a stub-tooth gear is $^{8}/_{10}$?

8) Explain how to determine the diametral pitch of a spur gear from a sample gear.

9) If a sample gear is cut to circular pitch, how can this pitch be determined?

10) What gage is used for seamless tubing, and does it apply to all metals?

11) How does the strength of iron wire rope compare with steel rope?

12) Is the friction between two bearing surfaces proportional to the pressure?

13) If the surfaces are well lubricated, upon what does frictional resistance depend?

14) What is the general rule for subtracting a negative number from a positive number? For example, $8 - (-4) = ?$

15) Is 1 meter longer than 1 yard?

16) On *Handbook* page **2874**, two of the equivalents of horsepower-hour are: 1,980,000 foot-pounds and 2.64 pounds of water evaporated at 212° F. How is this relationship between work and heat established?

17) Are "extra strong" and "double extra strong" wrought or steel pipe larger in diameter than standard-weight pipe?

18) In the design of plain bearings, what is the general relationship between surface finish and hardness of journal?

19) Are the nominal sizes of wrought or steel pipe ever designated by giving the outside diameter?

20) What are the advantages of plastics pipe?

21) Will charcoal ignite at a lower temperature than dry pine?

22) What general classes of steel are referred to as "stainless"?

23) What are free-cutting steels?

24) Does the nominal length of a file include the tang? For example, is a 12-inch file 12 inches long overall?

25) Is steel heavier (denser) than cast iron?

26) What is meant by specific heat?

27) What is the specific gravity (a) of solid bodies, (b) of liquids, (c) of gases?

28) A system of four-digit designations for wrought aluminum and aluminum alloys was adopted by The Aluminum Association in 1954. What do the various digits signify?

29) What alloys are known as "red brass," and how do they compare with "yellow brass"?

30) What is the difference between adiabatic expansion or compression and isothermal expansion or compression?

31) Are the sizes of all small twist drills designated by numbers?

32) Why are steel tools frequently heated in molten baths to harden them?

33) In hardening tool steel, what is the best temperature for refining the grain of the steel?

34) In cutting a screw thread on a tap, assume that the pitch is to be increased from 0.125 inch to 0.1255 inch to compensate for shrinkage in hardening. How can this be done?

35) What is the general rule for reading a vernier scale (a) for linear measurements; (b) for angular measurements?

36) The end of a shaft is to be turned to a taper of $\frac{3}{8}$ inch per foot for a length of inches without leaving a shoulder at the end of the cut. How is the diameter of the small end determined?

37) Is there a simple way of converting the function of 90° plus an angle to the function of the angle itself?

38) What decimal part of a degree is 53 minutes?

39) If $10x - 5 = 3x + 16$, what is the value of x?

40) Approximately what angle is required for a cone clutch to prevent either slipping or excessive wedging action?

41) What is the coefficient of friction?

42) Is Stub's steel wire gage used for the same purpose as Stub's iron wire gage?

43) Why are some ratchet mechanisms equipped with two pawls of different lengths?

44) How does the modulus of elasticity affect the application of flat belts?

45) What is the effect of centrifugal force on flat and V-belts?

46) Is the ultimate strength of a crane or hoisting chain equal to twice the ultimate strength of the bar or rod used for making the links?

47) How would you determine the size of chain required for lifting a given weight?

48) If a shaft $3\frac{1}{2}$ inches in diameter is to be turned at a cutting speed of 90 feet per minute, what number of revolutions per minute will be required?

49) In lapping by the "wet method," what kind of lubricant is preferable (a) with a steel lap, (b) with a cast-iron lap?

50) What is the meaning of the terms "right-hand" and "left-hand" as applied to helical or spiral gears, and how is the "hand" of the gear determined?

51) Are mating helical or spiral gears always made to the same hand?

52) How would you determine the total weight of 100 feet of $1\frac{1}{2}$-inch standard-weight pipe?

53) What is the difference between casehardening and pack-hardening?

54) What is the nitriding process of heat-treating steel?

55) What is the difference between single-cut and double-cut files?

56) For general purposes, what is the usual height of work benches?

57) What do the terms "major diameter" and "minor diameter" mean as applied to screw threads in connection with the American Standard?

58) Is the present SAE Standard for screw threads the same as the Unified and American Standard?

59) Does the machinability of steel depend only upon its hardness?

60) Is there any direct relationship between the hardness of steel and its strength?

61) What is the millimeter equivalent of $^{33}/_{64}$ of an inch?

62) How is the sevolute function of an angle calculated?

63) What is the recommended cutting speed in feet per minute for turning normalized AISI 4320 alloy steel with a hardness of 250 BHN, when using a coated, tough carbide tool?

64) The diametral pitch of a spur gear equals the number of teeth divided by pitch diameter. Is the diametral pitch of the cutter or hob for a helical or spiral gear determined in the same way?

65) Why are casehardening steels preferred for some gears and what special heat treatment is recommended?

66) Are the symbols for dimensions and angles used in spline calculations the same for both inch-dimension and metric module involute splines?

67) What kind of bearing surface and tool insert rake are provided by an indexable insert tool holder?

68) Is it necessary in making ordinary working drawings of gears to lay out the tooth curves? Why?

69) In milling plate cams on a milling machine, how is the cam rise varied other than by changing the gears between the dividing head and feed screw?

70) How is the angle of the dividing head spindle determined for milling plate cams?

71) How is the center-to-center distance between two gears determined if the number of teeth and diametral pitch are known?

72) How is the center-to-center distance determined for internal gears?

73) In the failure of riveted joints, rivets may fail through one or two cross-sections or by crushing. How may plates fail?

74) What gage is used in Britain to designate wire sizes?

75) What is a transmission dynamometer?

76) What is the advantage of a dynamometer for measuring power?

77) If a beam supported at each end is uniformly loaded throughout its length, will its load capacity exceed that of a similar beam loaded at the center only?

78) Is there any relationship between Brinell hardness and tensile strength of steel?

79) Is the outside diameter of a 2-inch pipe about 2 inches?

80) The hub of a lever 10 inches long is secured to a 1-inch shaft by a taper pin. If the maximum pull at the end of the lever equals 60 pounds, what pin diameter is required? (Give mean diameter or diameter at center.)

81) What are the two laws that form the basis of all formulas relating to the solution of triangles?

82) What are the sine and the cosine of the angle 45 degrees?

83) How is the pressure of water in pounds per square inch determined for any depth?

84) When calculating the basic load rating for a unit consisting of two bearings mounted in tandem, is the rated load of the combination equal to 2 times the capacity of a single bearing?

85) If a machine producing 50 parts per day is replaced by a machine that produces 100 parts per day, what is the percentage of increase?

86) If production is decreased from 100 to 50, what is the percentage of reduction?

87) What kind of steel is used ordinarily for springs in the automotive industry?

88) What is the heat-treating process known as “normalizing”?

89) What are the important Standards applicable to electric motors?

90) Is there an American Standard for section linings to represent different materials on drawings?

91) Is the taper per foot of the Morse Standard uniform for all numbers or sizes?

92) Is there more than one way to remove a tap that has broken in the hole during tapping?

93) The center-to-center distance between two bearings for gears is to be 10 inches, with a tolerance of 0.005 inch. Should this tolerance be (a) unilateral and plus, (b) unilateral and minus, (c) bilateral?

94) How are the available pitch-diameter tolerances for Acme screw threads obtained?

95) On *Handbook* page **1403**, there is a rule for determining the pressure required for punching circular holes into steel sheets or plates. Why is the product of the hole diameter and stock thickness multiplied by 80 to obtain the approximate pressure in tons?

96) What gage is used in the United States for cold-rolled sheet steel?

97) What gage is used for brass wire and is the same gage used for brass sheets?

98) Is the term "babbitt metal" applied to a single composition?

99) What are the chief elements in high-grade babbitt metal?

100) How many bars of stock 20 feet long will be needed to make 20,000 dowel-pins 2 inches long if the tool for cutting them off is 0.100 inch wide?

101) What is the melting point and density of cast iron; steel; lead; copper; nickel?

102) What lubricant is recommended for machining aluminum?

103) What relief angles are recommended for turning copper, brass, bronze, and aluminum?

104) Why is stock annealed between drawing operations in producing parts in drawing dies?

105) When is it advisable to mill screw threads?

106) How does a fluted chucking reamer differ from a rose chucking reamer?

107) What kind of material is commonly used for gage blocks?

108) What grade of gage blocks is used for shop standards?

109) What is the "lead" of a milling machine?

110) The table in *Machinery's Handbook 32 Digital Edition* page **3573** shows that a lead of 9.625 inches will be obtained if the numbers of teeth in the *driven* gears are 44 and 28 and the numbers of teeth on the *driving* gears 32 and 40. Prove that this lead of 9.625 inches is correct.

111) Use the prime number and factor table beginning on *Handbook* page **15** to reduce the following fractions to their lowest terms: $^{210}/_{462}$; $^{2765}/_{6405}$; $^{741}/_{1131}$.

112) If a bevel gear and a spur gear each have 30 teeth of 4 diametral pitch, how do the tooth sizes compare?

113) For what types of work are the following machinists' files used: (a) flat files? (b) half round files? (c) hand files? (d) knife files? (e) general-purpose files? (f) pillar files?

114) Referring to the illustration on *Handbook* page **697**, what is the dimension x over the rods used for measuring the dovetail slide if a is 4 inches, angle α is 60 degrees, and the diameter of the rods used is $^5/_8$ inch?

115) Determine the diameter of the bar or rod for making the links of a single chain required to lift safely a load of 6 tons.

116) Why will a helical gear have a greater tendency to slip on an arbor while the teeth are being milled than when milling a straight tooth gear?

117) What is meant by "trepanning"?

118) When is a removable or "slip" bushing used in a jig?

119) What are the relative ratings and properties of an H43 molybdenum high-speed tool steel?

120) What systematic procedure may be used in designing a roller chain drive to meet certain requirements as to horsepower, center distance, etc.?

121) In the solution of oblique triangles having two sides and the angle opposite one of the sides known, it is possible to have no solution or more than one solution. Under what condition will there be no solution?

122) What gear steels would you use (1) for casehardened gears? (2) for fully hardened gears? (3) for gears that are to be machined after heat treatment?

123) Is it practicable to tap holes and obtain (1) Class 2B fits? (2) Class 3B fits?

124) What is the maximum safe operating speed of an organic bonded type grinding wheel when used in a bench grinder?

125) What is the recommended type of diamond wheel and abrasive specification for internal grinding?

126) Is there a standard direction of rotation for all types of nonreversing electric motors?

127) What conditions are best suited for grease-lubricated antifriction bearings? What materials make up the grease used in this application?

128) In the example on *Handbook* page **2182**, the side relief angle at the leading edge of the single-point Acme thread cutting tool was calculated to be 19.27°, or 19°16′, which provides an effective relief angle (a_e) between the flank of the tool and the side of the thread of 10° at the minor diameter. What is the effective relief angle of this tool at the pitch diameter (E) and at the major diameter (D)? The pitch diameter of the thread is 0.900 inch, the major diameter is 1.000 inch, and the lead of the thread is 0.400 inch.

129) Helical flute milling cutters having eccentric relief are known to provide better support of the cutting edge than cutters ground with straight or concave relief. For a 1-inch-diameter milling cutter having a 35-degree helix angle, what is the measured indicator drop according to the methods described beginning on *Handbook* page **889** if the radial relief angle is to be 7°?

130) On *Handbook* page **2442**, **Table 4** shows that TFE fabric bearings have a load capacity of 60,000 pounds per square inch. Also shown in the table is a PV limit of 25,000 for this material. At what maximum surface speed in feet per minute can this material operate when the load is 60,000 pounds per square inch (psi)?

131) Is there a Standard for shaft diameter and housing bore tolerance limits that applies to rolling element bearings?

132) In designing an aluminum bronze plain bearing, what hardness should the steel journal have?

133) Steel balls are usually sold by the pound. How many pounds will provide 100 balls of $^{13}/_{32}$-inch-diameter carbon steel?

134) If a 3AM1-18 steel retaining ring were used on a rotating shaft, what is the maximum allowable speed of rotation?

135) What procedure applies to 3-wire measurements of Acme threads when the lead angle is greater than 5 degrees?

136) Twelve $1^1/_2$-inch-diameter rods are to be packed in a tube. What is the minimum inside diameter of the tube?

137) A four-wheel dolly supports a rack that holds six 5-gallon bottles filled with water that must be moved up a ramp onto the bed of a truck. What force is required to pull the loaded dolly up the ramp into the truck if the ramp is 14 feet long and the truck bed is 44 inches high? Neglect the weight of the dolly and rack, and friction in the dolly wheels. What if the ramp was 8 feet long?

SECTION 24

ANSWERS TO EXERCISES AND REVIEW QUESTIONS

All references are to *Machinery's Handbook, 32nd Edition* and *Machinery's Handbook 32 Digital Edition* page numbers

Answers to *Practice Exercises for Section 1 (see* **page 3***)*

Number of Exercise	Answers (Or where information is given in the *Handbook*)
1	78.54 mm^2; 31.416 mm
2	4.995 or 5, approx.
3	3141.6 mm^2
4	127.3 psi
5	1.27
6	1.5708
7	8 hours, 50 minutes
8	2450.448 pounds
9	$2\frac{1}{16}$ inches
10	7 degrees, 10 minutes
11	$(\pi - 2)r^2 = 1.1416r^2$
12	Formula on *Handbook 32 Digital Edition* page **3770**
13	740 gallons, approximately
14	Formula on *Handbook* page **86**
15	Formula on *Handbook* page **86**
16	Formula on *Handbook* page **86**

Answers to *Practice Exercises for Section 2 (see* **page 11***)*

Number of Exercise	Answers (Or where information is given in the *Handbook*)
1	*Handbook* pages **711** and **720**
2	(a) 0.043 inch, (b) 0.055 inch, (c) 0.102 inch
3	0.336 inch
4	2.796 inches
5	4.743 inches
6	4.221 feet
7	*Handbook* page **70** and **79**
8	$x = 2.585$, $y = -0.456$
9	$x = 2.878$, $y = 3.279$
10	$x = 1.629$, $y = 1.787$
11	$x = 1.000$, $y = 2.000$

Answers to *Practice Exercises for Section 3 (see* **page 25***)*

Number of Exercise	Answers (Or where information is given in the *Handbook*)
1	(a) 104 horsepower; (b) if reciprocal is used, $H = 0.33\ D^2 SN$
2	65 inches
3	5.74 inches
4	Side $s = 5.77$ inches; diagonal $d = 8.165$ inches, and volume = 192.1 cubic inches
5	91.0408 square inches
6	4.1888 and 0.5236
7	59.217 cubic inches
8	*Handbook* page **2838**
9	$a = \frac{2A}{h} - b$
10	$r = \sqrt{R^2 - \frac{s^2}{4}}$
11	$a = \sqrt{\frac{(P/\pi)^2}{2} - b^2}$
12	$\sin A = \sqrt{1 - \cos^2 A}$
13	$a = \frac{b \times \sin A}{\sin B}$; $b = \frac{a \times \sin B}{\sin A}$ $\sin A = \frac{a \times \sin B}{b}$ $\sin B = \frac{b \times \sin A}{a}$

Answers to *Practice Exercises for Section 4 (see* **page 36***)*

Number of Exercise	Answers (Or where information is given in the *Handbook*)
2	4; 35; 72
3	$A2 × B$1. The dollar sign ($) indicates an absolute reference to the row or column it precedes. For $A2, the referenced cell is in column A, and row changes are relative to starting row 2. For B$1, the referenced cell is in row 1, and the column changes relative to column B.
6	$5,954.45; $6,131.81

Answers to *Practice Exercises for Section 4 (Continued)*

Number of Exercise	Answers (Or where information is given in the *Handbook*)
7	−2; undefined; −16
8	No, operations with exponents are performed before multiplication, division, addition, or subtraction. Refer to spreadsheet documentation for operator precedence rules.

Answers to *Practice Exercises for Section 5 (see* **page 46***)*

Number of Exercise	Answers (Or where information is given in the *Handbook*)
1	*Guide* page **41**
2	Calculator, or table beginning on *Handbook 32 Digital Edition* page **3175**
3	2; 2; 1; $\overline{3}$; 3; 1
4	As location of decimal point is indicated by characteristic, which is not given, the number might be 7082, 708.20, 70.82, 7.082, 0.7082, 0.07082, etc.; 7675, 767.5, etc.; 1689, 168.9, etc.
5	(a) 70.82; 76.75; 16.89; (b) 708.2; 767.5, 168.9; 7.082, 7.675, 1.689; 7082, 7675, 1689
6	2.88389; 1.94052; $\overline{3}$.94151
7	792.4; 17.49; 1.514; 486.5
8	4.87614; 1.62363
9	67.603; 4.7547
10	146.17; 36.8
11	9.88; 5.422; 5.208
12	0.2783
13	0.0000001432
14	237.6
15	187.08
16	14.403 square inches
17	2.203 or, say, 2¼ inches
18	107 horsepower
19	No
20	Yes, see *Handbook* page **2187**

Answers to *Practice Exercises for Section 6 (see* **page 49***)*

Number of Exercise	Answers (Or where information is given in the *Handbook*)
1	8001.3 cubic inches
2	83.905 square inches
3	69.395 cubic inches
4	1.299 inches
5	22.516 cubic inches
6	8 inches
7	0.0276 cubic inch
8	4.2358 inches
9	1.9635 cubic inches
10	410.5024 cubic inches
11	26.4501 square inches
12	Radius, 1.4142 inches; area, 0.43 square inch
13	Area, 19.869 square feet; volume, 10.2102 cubic feet
14	Area, 240 square feet; volume, 277.12 cubic feet
15	11.3137 inches
16	41.03 gallons
17	17.872 square gallons
18	1.032 inches
19	40 cubic inches
20	Table on *Handbook* page **82**
21	Table on *Handbook* page **82**
22	5.0801 inches
23	4 inches; 5226 inches

Answers to *Practice Exercises for Section 7 (see* **page 54***)*

Number of Exercise	Answers (Or where information is given in the *Handbook*)
1	*Handbook* page **56**
2	*Handbook* page **56**
3	*Handbook* page **56**
4	*Handbook* page **56**
5	*Handbook* page **57**
6	*Handbook* page **57**
7	*Handbook* page **57**
8	*Handbook* page **57**
9	*Handbook* page **57**
10	*Handbook* page **57**
11	*Handbook* page **58**
12	*Handbook* page **58**

Answers to *Practice Exercises for Section 7 (Continued)*

Number of Exercise	Answers (Or where information is given in the *Handbook*)
13	*Handbook* page **58**
14	*Handbook* page **57**
15	*Handbook* page **58**
16	*Handbook* page **58**
17	*Handbook* page **58**
18	*Handbook* page **59**
19	*Handbook* page **59**
20	*Handbook* page **59**
21	*Handbook* page **59**
22	*Handbook* page **59**
23	*Handbook* page **60**
24	*Handbook* page **60**
25	*Handbook* page **60**
26	*Handbook* page **60**

Answers to *Practice Exercises for Section 8 (see* **page 63***)*

Number of Exercise	Answers (Or where information is given in the *Handbook*)
1	See *Handbook* pages **93–97**
2	In any right-angle triangle having an acute angle of 30 degrees, the side opposite that angle equals 0.5 × hypotenuse.
3	Sine = 0.31634; tangent = 0.51549; cosine = 0.83942
4	Angles equivalent to tangents are 27°29′24″ and 7°25′16″; angles equivalent to cosines are 86°5′8″ and 48°26′52″
5	Rule 1: Side opposite = hypotenuse × sine; Rule 2: Side opposite = side adjacent × tangent
6	Rule 1: Side adjacent = hypotenuse × cosine; Rule 2: Side adjacent = side opposite × cotangent
7	*Handbook* page **89**
8	*Handbook* page **91**

Answers to *Practice Exercises for Section 8 (Continued)*

Number of Exercise	Answers (Or where information is given in the *Handbook*)
9	Yes. This can be done after dividing the isosceles triangle into two right-angle triangles.

Answers to *Practice Exercises for Section 9 (see* **page 81***)*

Number of Exercise	Answers (Or where information is given in the *Handbook*)
1	2 degrees, 58 minutes
2	1 degree, 47 minutes
3	2.296 inches, as shown by the table on *Handbook* page **720**
4	$360°/N - 2a$ = angle intercepted by width W. The sine of $\frac{1}{2}$ this angle; $\frac{1}{2}B = \frac{1}{2}W$ hence, this sine $\times B = W$
5	3.1247 inches
6	3.5085 inches
7	1.7677 inches
8	75 feet approximately
9	$a = 1.0316$ inches; $b = 3.5540$ inches; $c = 2.2845$ inches; $d = 2.7225$ inches
10	$a = 18°22'$. For solution of similar problem, see *Guide*, **Example 4** of Section 8, page **60**
11	$A = 5.8758''$; $B = 6.0352''$; $C = 6.2851''$; $D = 6.4378''$; $E = 6.1549''$; $F = 5.8127''$. apply formula on *Handbook* page **95**
12	2°37′33″; 5°15′6″
13	5.2805 inches
14	10 degrees, 23 minutes

Answers to *Practice Exercises for Section 10 (see* **page 92***)*

Number of Exercise	Answers (Or where information is given in the *Handbook*)
1	84°; 63°31′; 32°29′
2	$B = 29°$; $b = 3.222$ feet; $c = 6.355$ feet; area = 10.013 square feet
3	$C = 22°$; $b = 2.33$ inches; $c = 1.358$ inches; area = 1.396 square inches
4	$A = 120°10′$; $a = 0.445$ foot; $c = 0.211$ foot; area = 0.027 square feet
5	The area of a triangle equals one-half the product of two of its sides multiplied by the sine of the angle between them. The area of a triangle may also be found by taking one-half of the product of the base and the altitude.

Answers to *Practice Exercises for Section 11 (see* **page 98***)*

Number of Exercise	Answers (Or where information is given in the *Handbook*)
1	*Handbook* page **1001** for Morse *Handbook* page **1012** for Jarno *Handbook* page **1012** for milling machine *Handbook* page **1893** for taper pins
2	2.205 inches; 12.694 inches
3	4.815 inches. *Handbook* page **701** $C = \frac{D-d}{2} \times \frac{\sqrt{1+\frac{T}{24}^2}}{\frac{T}{24}}$ $= \frac{2-1.75}{2} \times \frac{\sqrt{1+\left(\frac{0.62326}{24}\right)^2}}{\frac{0.62326}{24}}$
4	1.289 inches. *Handbook* page **702**

Answers to *Practice Exercises for Section 11 (Continued)*

Number of Exercise	Answers (Or where information is given in the *Handbook*)
5	3.110 inches. *Handbook* page **695**
6	0.0187 inch
7	0.2796 inch
8	1.000 inch
9	26 degrees, 7 minutes

Answers to *Practice Exercises for Section 12 (see* **page 112***)*

Number of Exercise	Answers (Or where information is given in the *Handbook*)
1	*Handbook* pages **643**, **645**
2	*Handbook* page **642**
3	*Handbook* page **642**
4	*Handbook* page **641**
5	*Handbook* page **674**
6	*Handbook* page **1950**
7	*Handbook* pages **1943**, **1954**
8	When the tolerance is unilateral
9	It means that a tolerance of 0.0004 to 0.0012 inch could normally be worked to. See table on *Handbook* page **649**
10	Yes. See *Handbook* page **802**

Answers to *Practice Exercises for Section 13 (see* **page 118**)

Number of Exercise	Answers (Or where information is given in the *Handbook*)
1	4000 pounds. *Handbook* page **2720**
2	*Handbook* page **2725**
3	430 balls. *Handbook 32 Digital Edition* page **3714**. To calculate, use density ρ from *Handbook* page **371** $$\frac{\text{balls}}{\text{lb}} = \frac{1}{\rho V} = \frac{6}{\rho \pi d^3}$$
4	¼ inch. *Handbook* page **2557**
5	0.172 inch. *Handbook* page **1936**
6	0.1251 to 0.1252. *Handbook* page **1885**
7	24,000 rpm. *Handbook* page **1904**
8	0.128 inch. *Handbook* page **1846**

Answers to *Practice Exercises for Section 14 (see* **page 125**)

Number of Exercise	Answers (Or where information is given in the *Handbook*)
1	Both countries have used the Unified Standard, but Britain is changing to the ISO Metric. See *Handbook* page **1943** and page **2047**
2	The symbol is used to specify an American Standard screw thread 3 inches in diameter, 4 threads per inch or the coarse series, and Class 2 fit.
3	An Acme thread is stronger, easier to cut with a die, and more readily engaged by a split nut used with a lead screw.
4	The Stub Acme form of thread is preferred for those applications where a coarse thread of shallow depth is required.
5	See tables, *Handbook* pages **1981–1982**
6	¾ inch per foot measured on the diameter-American and British Standards

Answers to *Practice Exercises for Section 14 (Continued)*

Number of Exercise	Answers (Or where information is given in the *Handbook*)
7	*Handbook* page **2096**
8	Center line of tool is set square to axis of screw thread.
9	Present practice is to set center line of tool square to axis of pipe.
10	See formulas for F_{rn} and F_{rs}, *Handbook* page **2068**
11	By three-wire method or by use of special micrometers. See *Handbook* pages **2125–2146**
12	Two quantities to be multiplied are the same as if enclosed by parentheses. See instructions about order of operations, *Handbook* page **4**
13	(a) Lead of double thread equals twice the pitch; (b) lead of triple thread equals three times the pitch. See *Handbook* page **2125**
14	See *Handbook* page **1952**
15	0.8337 inch. See *Handbook* page **2133**
16	No. Bulk of production is made to American Standard dimensions given in *Handbook*.
17	This Standard has been superseded by the American Standard.
18	Most machine screws (about 80% of the production) have the coarse series of pitches.
19	(a) Length includes head; (b) Length does not include head
20	No. 25. See table, *Handbook* page **2171**
21	0.1935 inch. See table, *Handbook* page **921**
22	Yes. The diameters decrease as the numbers increase.
23	The numbered sizes range in diameter from 0.0059 to 0.228 inch, and the letter sizes from 0.234 to 0.413 inch. See *Handbook* pages **919–929**

Answers to *Practice Exercises for Section 14 (Continued)*

Number of Exercise	Answers (Or where information is given in the *Handbook*)
24	A thread of $^3/_4$ standard depth has sufficient strength, and tap breakage is reduced.
25	(a) and (b) the American Standard Unified form
26	Cap-screws are made in the same pitches as the Coarse-, Fine-, and 8- thread series of the American Standard, class 2A.
27	For thread form, see *Handbook* page **2104**. There are seven standard diameters as shown on page **2105**.
28	*Handbook* page **957**
29	*Handbook* page **957**
30	0.90 × pitch. See *Handbook* page **2128**
31	To reduce errors in the finished thread
32	Included angle is 82° for each

Answers to *Practice Exercises for Section 15 (see* **page 141***)*

Number of Exercise	Answers (Or where information is given in the *Handbook*)
1	A foot-pound in mechanics is a unit of work and is the work equivalent to raising 1 pound 1 foot high.
2	1000 foot-pounds
3	Only as an average value. See *Handbook* page **184**
4	28 foot-pounds. See *Handbook* pages **182** and **184**
5	1346 pounds
6	Neglecting air resistance, the muzzle velocity is the same as the velocity with which the projectile strikes the ground. See *Handbook* page **176**
7	See *Handbook* page **157**

Answers to *Practice Exercises for Section 15 (Continued)*

Number of Exercise	Answers (Or where information is given in the *Handbook*)
8	Square
9	1843 pounds approximately
10	The pull will have been increased from 1843 pounds to about 2617 pounds. See *Handbook* page **170**
11	Yes
12	About 11 degrees
13	The angle of repose
14	The coefficient of friction equals the tangent of the angle of repose.
15	32.16 feet per second^2
16	No. 32.16 feet per second^2 is the value at sea level at a latitude of about 40 degrees, but this figure is commonly used. See *Handbook* page **152**
17	No. The rim stress is independent of the diameter and depends upon the velocity.
18	10 to 13. See *Handbook 32 Digital Edition* page **3236**
19	No. The increase in stress is proportional to the square of the rim velocity.
20	110 feet per second or approximately 1.25 miles per minute
21	Because the strength of wood is greater in proportion to its weight than cast iron.
22	See *Handbook* page **198**
23	In radians per second
24	A radian equals the angle subtended by the arc of circle; this angle is 57.3 degrees nearly.
25	*Handbook* page **99**
26	60 degrees; 72 degrees
27	*Handbook* page **99**

Answers to *Practice Exercises for Section 15 (Continued)*

Number of Exercise	Answers (Or where information is given in the *Handbook*)
28	*Handbook* page **177** (see *Guide* page **141** for example illustrating method of using tables)
29	Length of arc = radians × radius. As radius = 1 in the table segments, l = radians
30	40 degrees, 37 minutes, 42 seconds
31	176 radians per second; 1680.7 revolutions per minute
32	1.5705 inches
33	27.225 inches

Answers to *Practice Exercises for Section 16 (see* **page 157***)*

Number of Exercise	Answers (Or where information is given in the *Handbook*)
1	*Handbook* page **205**
2	12,000 pounds
3	1 inch
4	*Handbook* page **508**
5	*Handbook* page **201**
6	*Handbook* page **508**
7	3-inch diameter. See *Handbook* page **280**
8	Linearly
9	Plastics under high loads can creep, or slowly deform, due to their viscous nature. At higher temperatures, plastics are likely to become more viscous, and creep is more rapid. See *Handbook* page **568**
10	Matrix and fiber reinforcements. The fiber reinforcements provide most of the tensile strength. See *Handbook* page **586**

Answers to *Practice Exercises for Section 17 (see* **page 165***)*

Number of Exercise	Answers (Or where information is given in the *Handbook*)
1	1.568 (See formula on *Handbook* page **246**)
2	1.571
3	6200 pounds per square inch approximately

Answers to *Practice Exercises for Section 17 (Continued)*

Number of Exercise	Answers (Or where information is given in the *Handbook*)
4	It depends upon the class of service. See *Handbook* page **293**
5	Tangential load = 550 pounds; twisting moment = 4400 inch-pounds
6	See formulas on *Handbook* page **292**
7	The head is useful for withdrawing the key, especially when it is not possible to drive against the inner end. See *Handbook* page **2560**
8	Key is segment-shaped and fits into circular keyseat. See *Handbook* pages **2563–2564**
9	These keys are inexpensive to make from round bar stock, and keyseats are easily formed by milling.
10	0.211 inch. See table, *Handbook* page **2568**

Answers to *Practice Exercises for Section 18 (see* **page 176***)*

Number of Exercise	Answers (Or where information is given in the *Handbook*)
1	See text and footnote on *Handbook* page **2353**; see also *Guide* page **117**
2	American Standard B92.1, *Handbook* pages **2333** and **2339**
3	See text, *Handbook* page **2333**
4	See text, *Handbook* page **2333**
5	See definitions, *Handbook* page **2335**
6	None. See text, *Handbook* page **2339**
7	Yes, a crowned spline permits small amount of misalignment. See *Handbook* page **2350**
8	The torque capacity of splines may be calculated using the formulas and charts on *Handbook* page **2347–2351**
9	*Handbook* page **2346**
10	The fillet radius permits heavier loading and effects greater fatigue resistance than flat roots through absence of stress raisers.

Answers to *Practice Exercises for Section 19* (*see* **page 201**)

Number of Exercise	Answers (Or where information is given in the *Handbook*)
1	18 teeth; 3 inches; 0.2618 inch
2	2.666 inches; 2.333 inches; 0.166 inches
3	*Handbook* pages **2216–2217**
4	Chordal thickness at intersections of pitch circle with sides of tooth
5	**Table 3**, *Handbook* page **2215**
6	Calculate using **Table 5**, *Handbook* page **2217**
7	Surface durability stress and tooth fillet tensile stress are the two principal factors to be found in determining the power-transmitting capacity of spur gears.
8	Because the tooth shape varies as the number of teeth is changed
9	No. One hob may be used for all tooth numbers, and the same applies to any generating process.
10	Stub
11	*Handbook* (see *Fellows Stub Tooth* on page **2218**)
12	*Handbook* page **2298**
13	*Handbook* page **2228**
14	*Handbook* page **2218**
15	See table on *Handbook* page **2217**
16	*Handbook* page **2187**
17	*Handbook* pages **2244–2249**
18	*Handbook* page **2299**
19	Yes, but accurate tooth form is obtained only by a generating process.
20	See paragraph on *Handbook* page **2268**
21	*Handbook* page **2262**
22	When the numbers of teeth in both the pinion and the gear are the same, the pitch angle being 45 degrees for each.

Answers to *Practice Exercises for Section 19 (Continued)*

Number of Exercise	Answers (Or where information is given in the *Handbook*)
23	The whole depth minus the clearance between the bottom of a tooth space and the end of a mating tooth = the working depth.
24	See *Handbook* page **2222**
25	See *Handbook* pages **2210** and **2212**
26	See *Handbook* page **2263**
27	Circular pitch of gear equals linear pitch of worm.
28	Helix angle or lead angle of worm is measured from a plane perpendicular to the axis; helix angle of a helical gear is measured from the axis.
29	These terms each have the same meaning.
30	To provide a grinding allowance and to increase hob life over repeated sharpening
31	See explanation beginning on *Handbook 32 Digital Edition* page **3677**
32	*Handbook 32 Digital Edition* page **3677**
33	*Handbook* page **2277**
34	Normal diameter pitch is commonly used.
35	Yes (see *Handbook* page **2277**), but the hobbing process is generally applied.
36	Pitch diameter
37	*Handbook* page **2291**

Answers to *Practice Exercises for Section 20* (*see* **page 209**)

Number of Exercise	Answers (Or where information is given in the *Handbook*)
1	AISI 1108 CD ½ in. dia. = 1008 rpm 12L13, 150–200 BHN : = 1192 rpm 1040, Hot Rolled : = 611 rpm 1040, 375–425 BHN : = 214 rpm 41L40, 200–250 BHN : = 718 rpm 4140, Hot Rolled : = 611 rpm O2, Tool Steel : = 535 rpm M2, Tool Steel : = 497 rpm AISI 1108 CD 4 in. dia. = 126 rpm 12L13, 150–200 BHN : = 149 rpm 1040, Hot Rolled : = 576 rpm 1040, 375–425 BHN : = 27 rpm 41L40, 200–250 BHN : = 90 rpm 4140, Hot Rolled : = 76 rpm O2, Tool Steel : = 67 rpm M2, Tool Steel : = 62 rpm
2	AISI 1330, 200 BHN : 153 rpm 201 Stainless Steel, CD : 345 rpm ASTM Class 50 Gray Cast Iron : 145 rpm 6A1-4V Titanium Alloy : 52 rpm Waspaloy : 20 rpm (V = 60 fpm)
3	1½ in. dia.: 157 fpm—OK 3 in. dia. : 314 fpm—OK 4 in. dia. : 419 fpm—Too Fast
4	764 rpm
5	840 rpm (V = 110 fpm)
6	Operation: 1: N = 167 rpm; f_m = 13 in./min. 2: N = 127 rpm; f_m = 2.0 in./min. 3: N = 744 rpm 4: N = 458 rpm

Answers to *Practice Exercises for Section 20 (Continued)*

Number of Exercise	Answers (Or where information is given in the *Handbook*)
7	Existing operation: $V = 59$ fpm (too fast) $f_t = 0.011$ in/tooth (too severe) Change to: $V = 40$ fpm $N = 50$ rpm $f_t = 0.006$ in/tooth; $f_m = 3.6$ in/min
8	Existing operation: V = 188 fpm (too slow) $f_t = 0.006$ in/tooth (too slow) Change to: $V = 375$ fpm $N = 120$ rpm $f_t = 0.012$ in/tooth; $f_m = 520$ in/min
9	$V = 414$ fpm, $P_c = 9.0$ hp, $P_m = 11.24$ hp
10	$V = 104$ fpm
11	$V = 205$ fpm ($Q_{max} = 8.55$ in^3/min.; $f_m = 10.5$ in/min; $N = 131$ rpm)
12	$\frac{1}{4}$ in: $T = 123$ lb; $M = 6.38$ in-lb; $P_m = 0.19$ hp $\frac{1}{2}$ in: $T = 574$ lb; $M = 68$ in-lb; $P_m = 1.0$ hp 1 in: $T = 2712$ lb; $M = 711$ in-lb; $P_m = 5.3$ hp 19 mm: $T = 7244$ N; $M = 37.12$ N-m; $P_m = 2.43$ kw
13	$T = 1473$ lb; $M = 655$ in-lb; $P_m = 4.9$ hp
14	*Handbook* page **1065**
15	*Handbook* page **1066**
16	*Handbook* page **1070**
17	*Handbook* page **1070**
18	*Handbook* pages **851** and **1067**
19	*Handbook* pages **1128** and **1130**
20	*Handbook* pages **1096** and **1099**
21	*Handbook* page **1096**

Answers to *Practice Exercises for Section 20 (Continued)*

Number of Exercise	Answers (Or where information is given in the *Handbook*)
22	*Handbook* page **1133**
23	*Handbook* pages **1133–1134**
24	*Handbook* page **1139**
25	*Handbook* page **1145**
26	*Handbook* page **1177**

Answers to *Practice Exercises for Section 21 (see* **page 219***)*

Number of Exercise	Answers (Or where information is given in the *Handbook*)
1	Manual programming (also known as G-code programming), computer-aided manufacturing (also known as CAM programming), shop-floor (also known as conversational programming). See *Handbook* pages **1341–1342**
2	All machine activities: tool selection, motion control of axes, spindle speed, cutting feed rates, coolant flow, and more. See *Handbook* page **1338**
3	The address "G" identifies a preparatory command. See *Handbook* page **1344**
4	(d) milling and turning. See *Handbook* pages **1338–1387**
5	(b) Inches per revolution in/rev or mm/rev. See *Handbook* page **1348**
6	(b) F. See *Handbook* pages **1348** and **1350**
7	(c) I, J, and K. See *Handbook* pages **1350–1351**
8	(b) G43. See *Handbook* pages **1352–1353**
9	(d) R. See *Handbook* pages **1350–1351**
10	Sequence number. See *Handbook* page **1344**
11	True. See *Handbook* pages **1347–1348**
12	Cutter radius offset is an adjustment parallel to one of the axes. Cutter compensation is an adjustment that is normal to the part, whether or not the adjustment is parallel to an axis. See *Handbook* pages **1353–1354**

Answers to *Practice Exercises for Section 21 (Continued)*

Number of Exercise	Answers (Or where information is given in the *Handbook*)
13	(a) G40. See *Handbook* page **1353**
14	G98 = Initial level. G99 = R-level (return level). See *Handbook* pages **1354–1359**
15	a. Spot-drilling cycle — 5. G82 b. Peck-drilling cycle (deep hole) — 6. G83 c. Boring cycle, spindle rotating on withdrawal at feed rate — 3. G85 d. Drilling cycle — 2. G81 e. Right-hand threading cycle — 4. G84 f. Fixed cycle cancellation — 1. G80 See *Handbook* pages **1354–1359**
16	False. See *Handbook* page **1365**
17	A G word is a preparatory code consisting of the address G and two digits used to tell the control system to accept the remainder of the block in the required way. See *Handbook* page **1344**
18	The "right-hand rule" says that if a right-hand is laid palm up on the table of a vertical milling machine, the thumb will point in the positive *X* direction, the forefinger in the positive *Y* direction, and the erect middle finger in the positive *Z* direction. See *Handbook* pages **1374–1375**
19	True. See *Handbook* page **1383**
20	The machine type (mill, lathe, etc.), stock/raw material dimensions, part material, specific data related to the cutting tool, and work holding considerations. See *Handbook* page **1384**

Answers to *General Review Questions (see* **page 227***)*

Number of Exercise	Answers (Or where information is given in the *Handbook*)
1	*Handbook* page **77** gives the formula for length of side S in terms of the given area A.
2	The diameter of each end and the length of the taper; see explanation on *Handbook* page **1002**, also the table on page **1012**
3	Tolerance is applied in whatever direction is likely to be the least harmful; see *Handbook* page **643**
4	It is said that James Watt found, by experiment, that an average carthorse can develop 22,000 foot-pounds per minute, and added 50 percent to ensure good measure to purchasers of his engines ($22,000 \times 1.50 = 33,000$).
5	Tin in the high grades, and lead in the lower grades
6	Same depth as ordinary gear of 10 diametral pitch
7	The tooth thickness and the number of teeth are the same as an ordinary gear of 8 diametral pitch.
8	Add 2 to the number of teeth and divide by the outside diameter.
9	Multiply the outside diameter by 3.1416 and divide the product by the number of teeth plus 2.
10	Birmingham or Stub's iron wire gage is used for seamless steel, brass, copper, and aluminium tubing.
11	Iron wire rope has the least strength of all wire rope materials
12	If surfaces are well-lubricated, the friction is almost independent of the pressure, but if the surfaces are unlubricated, the friction is directly proportional to the normal pressure except for the higher pressures.
13	It depends very largely upon temperature. See *Handbook* section, *Lubricated Surfaces* on page **166**

Answers to *General Review Questions (Continued)*

Number of Exercise	Answers (Or where information is given in the *Handbook*)
14	8 – (–4) = 12. See rules for positive and negative numbers, *Handbook* page **3**
15	Yes. One meter equals 3.2808 feet; see other equivalents on *Handbook* page **2845**
16	Experiments have shown that there is a definite relationship between heat and work and that 1 British thermal unit equals 778 foot-pounds. To change 1 pound of water at 212° F into steam at that temperature requires about 966 British thermal units, or 966 × 788 = about 751,600 foot-pounds; hence, the number of pounds of water evaporated 212° F, equivalent to 1 horsepower-hour = 1,980,000 ÷ 751,600 = 2.64 pounds of water as given in *Handbook*, page **2874**
17	No. The thickness of the pipe is increased by reducing the inside diameter; compare thickness in the table on *Handbook 32 Digital Edition* page **3746**
18	As a general rule, smoother finishes are required for harder materials, for high loads, and for high speeds. See *Handbook* page **2402**
19	Yes. The so-called "O.D. pipe" begins, usually, with the $\frac{1}{4}$-inch size. See *Handbook 32 Digital Edition* page **3751**
20	It is light in weight and resists deterioration from corrosive or caustic fluids. See *Handbook 32 Digital Edition* page **3751**
21	Yes. About 140 degrees lower. See *Handbook* page **371**
22	Low-carbon alloy steels with high chromium content. See *Handbook* page **392**
23	Low-carbon steels containing 0.20% sulfur or less and usually from 0.90 to 1.20% manganese. See *Handbook* page **407**

Answers to *General Review Questions (Continued)*

Number of Exercise	Answers (Or where information is given in the *Handbook*)
24	No. The nominal length of a file indicates the distance from the point to the "heel" and does not include the tang.
25	Yes. See table, *Handbook* page **375**
26	Specific heat is a ratio of the amount of heat required to raise the temperature of a certain weight of substance 1° F to the amount of heat required to raise the temperature of an equivalent of water 1° F. See *Handbook* page **367**
27	(a) and (b) A number indicating how a given volume of the material or liquid compares in weight with an equal volume of water. (c) A number indicating a comparison in weight with an equal volume of air. See *Handbook* pages **375–376**
28	The first digit identifies the alloy type; the second, the impurity control; etc. See *Handbook* page **529**
29	Red brass contains 84 to 86% copper, about 5% tin, 5% lead, and 5% zinc, whereas yellow brass contains 62 to 67% copper, about 30% zinc, 1.5 to 3.5% lead, and not even 1% tin. See UNS Designations on *Handbook* pages **510**, **525**
30	See *Handbook 32 Digital Edition* pages **3261** and **3264**
31	No. Twenty-six sizes ranging from 0.234 to 0.413 inch are indicated by capital letters of the alphabet (see table, *Handbook* page **921–928**). Fractional sizes are also listed in manufacturers' catalogues beginning either at $\frac{1}{32}$ inch , $\frac{1}{16}$ inch, or $\frac{1}{8}$ inch, the smallest size varying with different firms.
32	To ensure uniform heating at a given temperature and protect the steel against oxidation. See *Handbook* page **469**

Answers to *General Review Questions (Continued)*

Number of Exercise	Answers (Or where information is given in the *Handbook*)
33	Hardening temperatures vary for different steels; see critical temperatures and how they are determined, *Handbook* pages **468–469**
34	Set the taper attachment to an angle the cosine of which equals 0.125 ÷ 0.1255. See *Handbook* page **2201**
35	See *Handbook* page **691**
36	Divide $\frac{3}{4}$ by 12; multiply the taper per inch found by 5 and subtract the result from the large diameter. See rules for figuring tapers, *Handbook* page **697**
37	Yes. See "Useful Relationships Among Angles," *Handbook* page **102**
38	0.8833. See *Handbook* page **99**
39	$x = 3$
40	About $12\frac{1}{2}$ degrees. See *Handbook* page **2533**
41	Ratio between resistance to the motion of a body due to friction, and the perpendicular pressure between the sliding and fixed surfaces. See formula, *Handbook* page **166**
42	No. Stub's steel wire gage applies to tool steel rod and wire, and the most important applications of Stub's iron wire gage (also known as Birmingham) are to seamless tubing, steel strips, and telephone and telegraph wire.
43	If the difference between the length of the pawls equals one-half of the pitch of the ratchet wheel teeth, the practical effect is that of reducing the pitch of one-half. See *Ratchet Gearing* starting on *Handbook* page **2296**
44	The high modulus of elasticity eliminates the need for periodic retensioning that is normally required with V-belts. See *Handbook* page **299**

Answers to *General Review Questions (Continued)*

Number of Exercise	Answers (Or where information is given in the *Handbook*)
45	Increasing centrifugal force has less effect on flat belts because of the low center of gravity. See *Handbook Flat Belting* starting on page **2572**
46	The ultimate strength is less due to bending action. See *Handbook 32 Digital Edition* page **3793**, and also see table, *Close-Link Hoisting, Sling and Crane Chain* on page **3797**
47	Refer to *Handbook 32 Digital Edition* page **3797**
48	Multiply 90 by 12 and divide the circumference of the shaft to obtain rpm. See cutting speed calculations, *Handbook* pages **1072–1074**
49	(a) Lard oil; (b) gasoline
50	If the teeth advance around the gear to the right, as viewed from one end, the gear is right-handed; and, if they advance to the left, it is a left-hand gear. See illustrations, *Handbook* page **2276**
51	No. They may be opposite hand depending upon the helix angle. See *Handbook* pages **2276–2277**
52	Multiply the total length by the weight per foot for plain end and coupled pipe, given in the table on *Handbook* page **2773**
53	The processes are similar but the term *pack-hardening* usually is applied to the casehardening of tool steel. See *Handbook* pages **479–480**
54	A gas process of surface hardening. See *Handbook* page **480**
55	See definitions for these terms given on *Handbook* page **1039**

Answers to *General Review Questions (Continued)*

Number of Exercise	Answers (Or where information is given in the *Handbook*)
56	About 34 inches, but the height may vary from 32 to 36 inches for heavy and light assembling, respectively.
57	Major diameter is the same as outside diameter, and the minor diameter is the same as root diameter. See definitions, on *Handbook* page **1947**
58	The SAE Standards conform, in general, with the Unified and American Standard Screw Thread Series as revised in 1959 and may, therefore, be considered to be virtually the same.
59	See information on work materials, *Handbook* page **1065**
60	Yes. See *Handbook* page **467** and page **505**
61	13.097 millimeters. See the table on *Handbook* page **2848**, which give millimeter equivalents of inch fractions and decimals.
62	The sevolute of an angle is obtained by subtracting the involute of the angle from the secant of that angle. See *Handbook* page **106**. The involute functions of angles are found in the tables beginning on *Handbook* page **107**
63	In **Table 1** starting on *Handbook* page **1083**, two feed-speed pairs are given, *Opt*. 28 in/tooth, 685 ft/min, and *Avg*., 13 in/tooth, 960 ft/min. These feed-speed pairs represent values for optimum and average conditions respectively and are intended as a starting point and generally require adjustment for actual cutting conditions. It is important to understand the factors that affect the cutting speed and feed, as covered in the footnote to **Table 1**, and *How to Use the Tables* which begins on *Handbook* page **1078**

Answers to *General Review Questions (Continued)*

Number of Exercise	Answers (Or where information is given in the *Handbook*)
64	No. First determine the diametral pitch the same as for a spur gear; then divide this "real diametral pitch" by the cosine of "real diametral pitch" by the cosine of the helix angle to obtain the "normal diametral pitch," which is the pitch of the cutter. See *Handbook* page **2277**
65	Casehardening steels can have hard, fine-grained surfaces and a soft, ductile core giving good strength combined with wear resistance. See *Handbook* page **2321**
66	Not in every instance. See *Handbook* page **2353**
67	A cemented carbide seat provides a flat bearing surface and a positive-, negative-, or neutral-rake orientation to the tool insert. See *Handbook* page **829**
68	No. The size of the gear blank, the pitch of the teeth, and depth of cut are sufficient for the operator in the shop. The tooth curvature is the result of the gear-cutting process. Tooth curves on the working drawing are of no practical value
69	By changing the inclination of the dividing head spindle. See *Handbook* page **2389**
70	See formula and example on *Handbook* page **2389**
71	Divide the total number of teeth in both gears by twice the diametral pitch to obtain the theoretical center-to-center distance. See formula in the table of *Formulas for Dimensions of Standard Spur Gears*, *Handbook* page **2212**
72	Subtract number of teeth on pinion from number of teeth on gear and divide the remainder by two times the diametral pitch. See Rule at bottom of *Handbook* page **2252**
73	See *Handbook* page **1865**

Answers to *General Review Questions (Continued)*

Number of Exercise	Answers (Or where information is given in the *Handbook*)
74	The Standard Wire Gage (S.W.G.), also known as the Imperial Wire Gage and as the English Legal Standard, is used in Britain for all wires.
75	A simple type of apparatus for measuring power
76	With a dynamometer, the actual amount of power delivered may be determined; that is, the power input minus losses. See *Handbook* page **2542**
77	The uniformly loaded beam has double the load capacity of a beam loaded at the center only. See formulas, *Handbook* page **251**
78	Refer to *Handbook* page **467** for graph of SAE-determined relationships.
79	No. The nominal size of steel pipe, except for sizes above 12 inches, is approximately equal to the inside diameter. See tables, *Handbook 32 Digital Edition* pages **3746** and **3748**
80	0.357 inch. See formula, *Handbook* page **221**
81	The laws of sines and cosines are stated on *Handbook* page **91**
82	Both the sine and cosine of 45 degrees are 0.70711.
83	Multiply depth in feet by 0.4335
84	No. See *Handbook* page **2492**
85	100%
86	50%
87	Various steels are used, depending on kind of spring. See *Handbook* page **403**
88	Normalizing is a special annealing process.The steel is heated above the critical range and allowed to cool in still air at ordinary temperature, *Handbook* page **479**. Normalizing temperatures for steels are given on *Handbook* pages **485–486**

Answers to *General Review Questions (Continued)*

Number of Exercise	Answers (Or where information is given in the *Handbook*)
89	The National Electrical Manufacturers Association (NEMA) Standards for mounting dimensions, frame sizes, horsepower, and speed ratings. See section beginning on *Handbook* page **2650**
90	Yes. The American Standard drafting room practice includes section lining, etc. See *Handbook* page **623**
91	No. There are different tapers per foot, ranging from 0.5986 to 0.6315 inch. See table on *Handbook* page **1001**
92	Yes. See *Handbook* page **2178**
93	Unilateral and plus. See *Handbook* page **642**
94	See table, *Handbook* page **2064**
95	If D = diameter of hole in inches; T = stock thickness in inches; shearing strength of steel = 51,000 pounds per square inch, then tonnage for punching $= 51{,}000 D\pi T/2000 = 80DT$
96	See *Handbook* pages **2712–2713**
97	The Brown & Sharpe or American wire gage is used for each. See *Handbook* pages **2708–2713**
98	No, this name is applied to several compositions that vary widely.
99	Antimony and copper
100	177 nearly. See table on *Handbook* page **1228**
101	See *Handbook* pages **362**, **371**, **375**
102	See *Handbook* page **1238**
103	See *Handbook* page **824**
104	See *Handbook 32 Digital Edition* page **3503**
105	See *Handbook* page **2201**

Answers to *General Review Questions (Continued)*

Number of Exercise	Answers (Or where information is given in the *Handbook*)
106	See *Handbook* page **897**
107	Steel, chromium-plated steel, chromium carbide, tungsten carbide, and other materials. See *Handbook* page **724**
108	See text on *Handbook* page **724**
109	The lead of a milling machine equals lead of helix or spiral milled when gears of equal size are placed on feed screw and wormgear stud; see rule for finding lead on *Handbook 32 Digital Edition* page **3567**
110	Multiply product of driven gears by lead of machine and divide by product of driving gears. If lead of machine is 10, divide 10 times product of driven gears by product of drivers.
111	$\frac{5}{11}$; $\frac{79}{183}$; $\frac{19}{29}$
112	The whole depth and tooth thickness at the large ends of the bevel-gear teeth are the same as the whole depth and thickness of spur-gear teeth of the same pitch.
113	See Text on *Handbook* page **1040**
114	5.7075 inches
115	Use the formula (found on *Handbook 32 Digital Edition* page **3793**) for finding the breaking load, which in this case is taken as three times the actual load. Transposing, $D = \sqrt{\frac{6 \times 2000 \times 3}{54,000}} = 0.816$, say, $\frac{7}{8}$ inch diameter
116	Because the direction of the cutter thrust tends to cause the gear to rotate upon the arbor. See *Handbook Milling the Helical Teeth* on page **2286**

Answers to *General Review Questions (Continued)*

Number of Exercise	Answers (Or where information is given in the *Handbook*)
117	Trepanning describes use of a fly-cutter or circular toothed cutter to cut a groove to the full depth of a plate, producing a hole of the required size. See *Handbook 32 Digital Edition* page **3488**
118	Chiefly when a hole is to be tapped or reamed after drilling. See *Handbook 32 Digital Edition* page **3389**
119	See table on *Handbook* page **445**
120	See *Handbook* page **2645**
121	See *Handbook* page **96**
122	See *Handbook* pages **2321–2322**
123	See *Handbook* page **985**
124	See table *Handbook* page **1301**
125	See table *Handbook* page **1285**
126	Motor rotation has been standardized by the National Electrical Manufacturers Association. See *Handbook* page **2652**
127	Grease-lubricated bearings are intended for slow-speed, moderate-to-high load applications. The grease is comprised of a mineral oil and a thickening agent, such as a metallic soap. See *Lubrication Selection* starting on *Handbook* page **2404**.

Answers to *General Review Questions (Continued)*

Number of Exercise	Answers (Or where information is given in the *Handbook*)
128	To solve this problem, the helix angle ϕ of the thread at the pitch and major diameters must be found, which is accomplished by substituting these diameters (E and D) for the minor diameters (K) in the formula for ϕ. Thus, at the pitch diameter: $\tan\phi = \frac{\text{lead of thread}}{\pi E} = \frac{0.400}{\pi \times 0.900}$ $\phi = 8.052° = 8°3'$ $a = a_e + \phi$ $a_e = a - \phi = 19°16' - 8°3' = 11°13'$ At the major diameter: $\tan\phi = \frac{\text{lead of thread}}{\pi D} = \frac{0.400}{\pi \times 1.000}$ $\phi = 7.256° = 7°15'$ $a_e = a - \phi = 19°16' - 7°15' = 12°1'$
129	0.0037 inch
130	$^5/_{12}$ foot (5 inches) per minute obtained by dividing 25,000 by 60,000. Note that this speed is considerably less than maximum surface speed at any load to prevent excess heat and wear
131	Yes. See **Table 14**, *Handbook* page **2463**, and following tables
132	550–600 BHN (Brinell Hardness Number) (See *Handbook* page **2402**)
133	1 pound. See *Handbook 32 Digital Edition* page **3714**
134	23,000 rpm. See *Handbook* page **1904**
135	See *Handbook* page **2137**
136	See footnote, **Table 2**, *Handbook 32 Digital Edition* page **3210**
137	See *Handbook* pages **2862** and **375**, and table on *Handbook* page **169**. Rope tension on 14-foot ramp = 64.3 lb; rope tension on 8-foot ramp= 111.1 lb

SECTION 25

UNITS OF MEASURE AND CONVERSION FACTORS

In the table of conversion factors that follows, the symbols for SI units, multiples, and submultiples are given in parentheses in the right-hand column. The symbol "*a*" following a number indicates that the conversion factor is exact.

Table 1. Metric Conversion Factors

Multiply	By	To Obtain
	Length	
centimeter	0.03280840	foot
centimeter	0.3937008	inch
fathom	1.8288[a]	meter (m)
foot	0.3048[a]	meter (m)
foot	30.48[a]	centimeter (cm)
foot	304.8[a]	millimeter (mm)
inch	0.0254[a]	meter (m)
inch	2.54[a]	centimeter (cm)
inch	25.4[a]	millimeter (mm)
kilometer	0.6213712	mile [US statute]
meter	39.37008	inch
meter	0.5468066	fathom
meter	3.280840	foot
meter	0.1988388	rod
meter	1.093613	yard
meter	0.0006213712	mile [US statute]
microinch	0.0254[a]	micrometer [micron] (μm)
micrometer [micron]	39.37008	microinch
mile [US statute]	1609.344[a]	meter (m)
mile [US statute]	1.609344[a]	kilometer (km)
millimeter	0.003280840	foot
millimeter	0.03937008	inch
rod	5.0292[a]	meter (m)
yard	0.9144[a]	meter (m)
	Area	
acre	4046.856	meter^2 (m^2)
acre	0.4046856	hectare
centimeter^2	0.1550003	inch^2
centimeter^2	0.001076391	foot^2
foot^2	0.09290304[a]	meter^2 (m^2)
foot^2	929.0304[a]	centimeter^2 (cm^2)
foot^2	92,903.04[a]	millimeter^2 (mm^2)

Table 1. Metric Conversion Factors *(Continued)*

Multiply	By	To Obtain
hectare	2.471054	acre
inch²	645.16[a]	millimeter² (mm²)
inch²	6.4516[a]	centimeter² (cm²)
inch²	0.00064516[a]	meter² (m²)
meter²	1550.003	inch²
meter²	10.763910	foot²
meter²	1.195990	yard²
meter²	0.0002471054	acre
mile²	2.5900	kilometer²
millimeter²	0.00001076391	foot²
millimeter²	0.001550003	inch²
yard²	0.8361274	meter² (m²)
Volume (including Capacity)		
centimeter³	0.06102376	inch³
foot³	28.31685	liter
foot³	28.31685	liter
gallon [UK liquid]	0.004546092	meter³ (m³)
gallon [UK liquid]	4.546092	liter
gallon [US liquid]	0.003785412	meter³ (m³)
gallon [US liquid]	3.785412	liter
inch³	16,387.06	millimeter³ (mm³)
inch³	16.38706	centimeter³ (cm³)
inch³	0.00001638706	meter³ (m³)
liter	0.001[a]	meter³ (m³)
liter	0.2199692	gallon [UK liquid]
liter	0.2641720	gallon [US liquid]
liter	0.03531466	foot³
meter³	219.9692	gallon [UK liquid]
meter³	264.1720	gallon [US liquid]
meter³	35.31466	foot³
meter³	1.307951	yard³
meter³	1000.[a]	liter
meter³	61,023.76	inch³
millimeter³	0.00006102376	inch³
quart [US liquid]	0.946	liter
quart [UK liquid]	1.136	liter
yard³	0.7645549	meter³ (m³)
Velocity, Acceleration, and Flow		
centimeter/second	1.968504	foot/minute
centimeter/second	0.03280840	foot/second
centimeter/minute	0.3937008	inch/minute
foot/hour	0.00008466667	meter/second (m/s)
foot/hour	0.00508[a]	meter/minute
foot/hour	0.3048[a]	meter/hour
foot/minute	0.508[a]	centimeter/second
foot/minute	18.288[a]	meter/hour
foot/minute	0.3048[a]	meter/minute
foot/minute	0.00508[a]	meter/second (m/s)

Table 1. Metric Conversion Factors *(Continued)*

Multiply	By	To Obtain
foot/second	30.48[a]	centimeter/second
foot/second	18.288[a]	meter/minute
foot/second	0.3048[a]	meter/second (m/s)
foot/second^2	0.3048[a]	meter/second^2 (m/s^2)
foot^3/minute	28.31685	liter/minute
foot^3/minute	0.0004719474	meter^3/second (m^3/s)
gallon [US liquid]/min.	0.003785412	meter^3/minute
gallon [US liquid]/min.	0.00006309020	meter^3/second (m^3/s)
gallon [US liquid]/min.	0.06309020	liter/second
gallon [US liquid]/min.	3.785412	liter/minute
gallon [UK liquid]/min.	0.004546092	meter^3/minute
gallon [UK liquid]/min.	0.00007576820	meter^3/second (m^3/s)
inch/minute	25.4[a]	millimeter/minute
inch/minute	2.54[a]	centimeter/minute
inch/minute	0.0254[a]	meter/minute
inch/second^2	0.0254[a]	meter/second^2 (m/s^2)
kilometer/hour	0.6213712	mile/hour [US statute]
liter/minute	0.03531466	foot^3/minute
liter/minute	0.2641720	gallon [US liquid]/minute
liter/second	15.85032	gallon [US liquid]/minute
mile/hour	1.609344[a]	kilometer/hour
millimeter/minute	0.03937008	inch/minute
meter/second	11,811.02	foot/hour
meter/second	196.8504	foot/minute
meter/second	3.280840	foot/second
meter/second^2	3.280840	foot/second^2
meter/second^2	39.37008	inch/second^2
meter/minute	3.280840	foot/minute
meter/minute	0.05468067	foot/second
meter/minute	39.37008	inch/minute
meter/hour	3.280840	foot/hour
meter/hour	0.05468067	foot/minute
meter^3/second	2118.880	foot^3/minute
meter^3/second	13,198.15	gallon [UK liquid]/minute
meter^3/second	15,850.32	gallon [US liquid]/minute
meter^3/minute	219.9692	gallon [UK liquid]/minute
meter^3/minute	264.1720	gallon [US liquid]/minute
	Mass and Density	
grain [$^1/_{7000}$ lb avoirdupois]	0.06479891	gram (g)
gram	15.43236	grain
gram	0.001[a]	kilogram (kg)
gram	0.03527397	ounce [avoirdupois]
gram	0.03215074	ounce [Troy]
gram/centimeter^3	0.03612730	pound/inch^3
hundredweight [long]	50.80235	kilogram (kg)
hundredweight [short]	45.35924	kilogram (kg)

Table 1. Metric Conversion Factors *(Continued)*

Multiply	By	To Obtain
kilogram	1000.[a]	gram (g)
kilogram	35.27397	ounce [avoirdupois]
kilogram	32.15074	ounce [Troy]
kilogram	2.204622	pound [avoirdupois]
kilogram	0.06852178	slug
kilogram	0.0009842064	ton [long]
kilogram	0.001102311	ton [short]
kilogram	0.001[a]	ton [metric]
kilogram	0.001[a]	tonne
kilogram	0.01968413	hundredweight [long]
kilogram	0.02204622	hundredweight [short]
kilogram/meter^3	0.06242797	pound/foot^3
kilogram/meter^3	0.01002242	pound/gallon [UK liquid]
kilogram/meter^3	0.008345406	pound/gallon [US liquid]
ounce [avoirdupois]	28.34952	gram (g)
ounce [avoirdupois]	0.02834952	kilogram (kg)
ounce [Troy]	31.10348	gram (g)
ounce [Troy]	0.03110348	kilogram (kg)
pound [avoirdupois]	0.4535924	kilogram (kg)
pound/foot^3	16.01846	kilogram/meter^3 (kg/m^3)
pound/inch^3	27.67990	gram/centimeter^3 (g/cm^3)
pound/gal [US liquid]	119.8264	kilogram/meter^3 (kg/m^3)
pound/gal [UK liquid]	99.77633	kilogram/meter^3 (kg/m^3)
slug	14.59390	kilogram (kg)
ton [long 2240 lb]	1016.047	kilogram (kg)
ton [short 2000 lb]	907.1847	kilogram (kg)
ton [metric]	1000.[a]	kilogram (kg)
ton [metric]	0.9842	ton [long 2240 lb]
ton [metric]	1.1023	ton [short 2000 lb]
tonne	1000.[a]	kilogram (kg)
	Force and Force/Length	
dyne	0.00001[a]	newton (N)
kilogram-force	9.806650[a]	newton (N)
kilopound	9.806650[a]	newton (N)
newton	0.1019716	kilogram-force
newton	0.1019716	kilopound
newton	0.2248089	pound-force
newton	100,000.[a]	dyne
newton	7.23301	poundal
newton	3.596942	ounce-force
newton/meter	0.005710148	pound/inch
newton/meter	0.06852178	pound/foot
ounce-force	0.2780139	newton (N)
pound-force	4.448222	newton (N)
poundal	0.1382550	newton (N)
pound/inch	175.1268	newton/meter (N/m)
pound/foot	14.59390	newton/meter (N/m)

Table 1. Metric Conversion Factors *(Continued)*

Multiply	By	To Obtain
Bending Moment or Torque		
dyne-centimeter	0.0000001[a]	newton-meter (N · m)
kilogram-meter	9.806650[a]	newton-meter (N · m)
ounce-inch	7.061552	newton-millimeter
ounce-inch	0.007061552	newton-meter (N · m)
newton-meter	0.7375621	pound-foot
newton-meter	10,000,000.[a]	dyne-centimeter
newton-meter	0.1019716	kilogram-meter
newton-meter	141.6119	ounce-inch
newton-millimeter	0.1416119	ounce-inch
pound-foot	1.355818	newton-meter (N · m)
Moment of Inertia and Section Modulus		
moment of inertia [kg · m²]	23.73036	pound-foot²
moment of inertia [kg · m²]	3417.171	pound-inch²
moment of inertia [lb · ft²]	0.04214011	kilogram-meter² (kg · m²)
moment of inertia [lb · inch²]	0.0002926397	kilogram-meter² (kg · m²)
moment of section [foot⁴]	0.008630975	meter⁴ (m⁴)
moment of section [inch⁴]	41.62314	centimeter⁴
moment of section [meter⁴]	115.8618	foot⁴
moment of section [cm⁴]	0.02402510	inch⁴
section modulus [foot³]	0.02831685	meter³ (m³)
section modulus [inch³]	0.00001638706	meter³ (m³)
section modulus [meter³]	35.31466	foot³
section modulus [meter³]	61,023.76	inch³
Momentum		
kilogram-meter/second	7.233011	pound-foot/second
kilogram-meter/second	86.79614	pound-inch/second
pound-foot/second	0.1382550	kilogram-meter/second (kg · m/s)
pound-inch/second	0.01152125	kilogram-meter/second (kg · m/s)
Pressure and Stress		
atmosphere [14.6959 lb/inch²]	101,325.	pascal (Pa)
bar	100,000.[a]	pascal (Pa)
bar	14.50377	pound/inch²
bar	100,000.[a]	newton/meter² (N/m²)
hectobar	0.6474898	ton [long]/inch²
kilogram/centimeter²	14.22334	pound/inch²
kilogram/meter²	9.806650[a]	newton/meter² (N/m²)
kilogram/meter²	9.806650[a]	pascal (Pa)
kilogram/meter²	0.2048161	pound/foot²
kilonewton/meter²	0.1450377	pound/inch²
newton/centimeter²	1.450377	pound/inch²
newton/meter²	0.00001[a]	bar
newton/meter²	1.0[a]	pascal (Pa)
newton/meter²	0.0001450377	pound/inch²
newton/meter²	0.1019716	kilogram/meter²
newton/millimeter²	145.0377	pound/inch²

Table 1. Metric Conversion Factors *(Continued)*

Multiply	By	To Obtain
pascal	0.00000986923	atmosphere
pascal	0.00001[a]	bar
pascal	0.1019716	kilogram/meter^2
pascal	1.0[a]	newton/meter^2 (N/m^2)
pascal	0.02088543	pound/foot^2
pascal	0.0001450377	pound/inch^2
pound/foot^2	4.882429	kilogram/meter^2
pound/foot^2	47.88026	pascal (Pa)
pound/inch^2	0.06894757	bar
pound/inch^2	0.07030697	kilogram/centimeter^2
pound/inch^2	0.6894757	newton/centimeter^2
pound/inch^2	6.894757	kilonewton/meter^2
pound/inch^2	6894.757	newton/meter^2 (N/m^2)
pound/inch^2	0.006894757	newton/millimeter^2 (N/mm^2)
pound/inch^2	6894.757	pascal (Pa)
ton [long]/inch^2	1.544426	hectobar
Energy and Work		
Btu [International Table]	1055.056	joule (J)
Btu [mean]	1055.87	joule (J)
calorie [mean]	4.19002	joule (J)
foot-pound	1.355818	joule (J)
foot-poundal	0.04214011	joule (J)
joule	0.0009478170	Btu [International Table]
joule	0.0009470863	Btu [mean]
joule	0.2386623	calorie [mean]
joule	0.7375621	foot-pound
joule	23.73036	foot-poundal
joule	0.9998180	joule [International US]
joule	0.9999830	joule [US legal, 1948]
joule [International US]	1.000182	joule (J)
joule [US legal, 1948]	1.000017	joule (J)
joule	.0002777778	watt-hour
watt-hour	3600.[a]	joule (J)
Power		
Btu [International Table]/hour	0.2930711	watt (W)
foot-pound/hour	0.0003766161	watt (W)
foot-pound/minute	0.02259697	watt (W)
horsepower [550 ft-lb/s]	0.7456999	kilowatt (kW)
horsepower [550 ft-lb/s]	745.6999	watt (W)
horsepower [electric]	746.[a]	watt (W)
horsepower [metric]	735.499	watt (W)
horsepower [UK]	745.70	watt (W)
kilowatt	1.341022	horsepower [550 ft-lb/s]
watt	2655.224	foot-pound/hour
watt	44.25372	foot-pound/minute
watt	0.001341022	horsepower [550 ft-lb/s]
watt	0.001340483	horsepower [electric]
watt	0.001359621	horsepower [metric]
watt	0.001341022	horsepower [UK]
watt	3.412141	Btu [International Table]/hour

Table 1. Metric Conversion Factors *(Continued)*

Multiply	By	To Obtain
	Viscosity	
poise	0.1[a]	pascal-second (Pa · s)
centipoise	0.001[a]	pascal-second (Pa · s)
stoke	0.0001[a]	meter²/second (m^2/s)
centistoke	0.000001[a]	meter²/second (m^2/s)
meter²/second	1,000,000.[a]	centistoke
meter²/second	10,000.[a]	stoke
pascal-second	1000.[a]	centipoise
pascal-second	10.[a]	poise
	Temperature	
temperature Celsius, t_C	temperature Kelvin, t_K	$t_K = t_C + 273.15$
temperature Fahrenheit, t_F	temperature Kelvin, t_K	$t_K = (t_F + 459.67)/1.8$
temperature Celsius, t_C	temperature Fahrenheit, t_F	$t_F = 1.8\, t_C + 32$
temperature Fahrenheit, t_F	temperature Celsius, t_C	$t_C = (t_F - 32)/1.8$
temperature Kelvin, t_K	temperature Celsius, t_C	$t_C = t_K - 273.15$
temperature Kelvin, t_K	temperature Fahrenheit, t_F	$t_F = 1.8\, t_K - 459.67$
temperature Kelvin, t_K	temperature Rankine, t_R	$t_R = 9/5\, t_K$
temperature Rankine, t_R	temperature Kelvin, t_K	$t_K = 5/9\, t_R$

[a] The figure is exact.

Table 2. Circular and Angular Measure Conversion Factors

Circumference of circle = 360 degrees = 2π radian = 6.283185 radian

1 *degree* (°) = 60 minutes = 3600 seconds = $\pi/180$ radian = 0.017453 radian

1 *quadrant* = 90 degrees = $\pi/2$ radians = 1.570796 radians

1 *minute* (′) = 60 seconds = 0.016667 degree = 0.000291 radian

1 *radian* = 57.2957795 degrees

$\pi = 3.141592654$

Table 3. Linear Measure Conversion Factors

Metric

1 *kilometer (km)* =
- **1000** meters
- **100,000** centimeters
- **1,000,000** millimeters
- 0.539956 nautical mile
- 0.621371 mile
- 1093.61 yards
- 3280.83 feet
- 39,370.08 inches

1 *meter (m)* =
- **10** decimeters
- **100** centimeters
- **1000** millimeters
- 1.09361 yards
- 3.28084 feet
- 39.37008 inches

1 *decimeter (dm)* = **10** *centimeters*

1 *centimeter (cm)* =
- **0.01** meter
- **10 millimeters**
- 0.0328 foot
- 0.3937 inch

1 *millimeter (mm)* =
- **0.001** meter
- **0.1** centimeter
- **1000** microns
- 0.03937 inch

1 *micrometer or micron (μm)* =
- **0.000001** meter = one millionth meter
- **0.0001** centimeter
- **0.001** millimeter
- 0.00003937 inch
- 39.37 micro-inches

US Customary

1 *mile (mi)* =
- 0.868976 nautical mile
- **1760** yards
- **5280** feet
- **63,360** inches
- **1.609344** kilometers
- **1609.344** meters
- **160,934.4** centimeters
- **1,609,344** millimeters

1 *yard (yd)* =
- **3** feet
- **36** inches
- **0.9144** meter
- **91.44** centimeter
- **914.4** millimeter

1 *foot (international) (ft)* =
- **12** inches = $\frac{1}{3}$ yard
- **0.3048** meter
- **30.48** centimeter
- **304.8** millimeters

1 *survey foot* =
- 1.000002 international feet
- $\frac{12}{39.37}$ = 0.3048006096012 meter

1 *inch (in)* =
- **1000** mils
- **1,000,000** micro-inches
- **2.54** centimeters
- **25.4** millimeters
- **25,400** microns

1 *mil* =
- **0.001** inch
- **1000** micro-inches
- **0.0254** millimeters

1 *micro-inch (μin)* =
- **0.000001** inch = one millionth inch
- **0.0254** micrometer (micron)

Surveyors Measure

1 *mile* = **8** furlongs = **80** chains

1 *furlong* = **10** chains = **220** yards

1 *chain* =
- **4** rods = **22** yards = **66** feet = **100** links

1 *rod* =
- **5.5** yards = **16.5** feet = **25** links
- 5.0292 meters

1 *link* = 7.92 inches

1 *span* = 9 inches

1 *hand* = 4 inches

Nautical Measure

1 *league* = 3 nautical miles

1 *nautical mile* =
- 1.1508 statute miles
- **6076.11549** feet
- 1.8516 kilometers

1 *fathom* = **2** yards = **6** feet

1 *knot* = nautical unit of speed =
- 1 nautical mile per hour
- 1.1508 statute miles per hour
- 1.8516 kilometers per hour

One degree at the equator =
- 60 nautical miles
- 69.047 statute miles
- 111.098 kilometers

One minute at the equator =
- 1 nautical mile
- 1.1508 statute miles
- 1.8516 kilometers

360 degrees at the equator =
- circumference at equator
- 21,600 nautical miles
- 24,856.8 statute miles
- 39,995.4 kilometers

Note: Figures in **Bold** indicate exact conversion values

Table 4. Square Measure and Conversion Factors

Metric System

1 *square kilometer* (km^2) =
- **100** hectares
- **1,000,000** square meters
- 0.3861 square mile
- 247.1 acres

1 *hectare (ha)* =
- **0.01** square kilometer
- **100 ares**
- **10,000** square meters
- 2.471 acres
- 107,639 square feet

1 *are (a)* =
- **0.0001** square kilometer
- **100** square meters
- 0.0247 acre
- 1076.4 square feet

1 *square meter* (m^2) =
- **0.000001** square kilometer
- **100** square decimeters
- **10000** square centimeters
- **1,000,000** square millimeters
- 10.764 square feet
- 1.196 square yards

1 *square decimeter* (dm^2) =
- **100** square centimeters

1 *square centimeter* (cm^2) =
- **0.0001** square meters
- **100** square millimeters
- **0.001076** square foot
- **0.155** square inch

1 *square millimeter* (mm^2) =
- **0.01** square centimeters
- **1,000,000** square microns
- 0.00155 square inch

1 *square micrometer (micron)* (μm^2) =
- $\mathbf{1 \times 10^{-12}}$ square meter
- **0.000001** square millimeters
- 1×10^{-9} square inch
- 1549.997 square micro-inch

US Customary System

1 *square mile* (mi^2) =
- **640** acres
- **6400** square chains
- 2.5899 square kilometers

1 *acre* =
- **10** square chains
- **4840** square yards
- **43,560** square feet
- a square, 208.71 feet on a side
- 0.4046856 hectare
- 40.47 ares
- 4046.856 square meters

1 *square chain* =
- **16** square rods
- **484** square yards
- **4356** square feet

1 *square rod* =
- **30.25** square yards
- **272.25** square feet
- **625** square links

1 *square yard* (yd^2) =
- **9** square feet
- **1296** square inches
- **0.83612736** square meter
- **8361.2736** square centimeter
- **836,127.36** square millimeter

1 *square foot* (ft^2) =
- 0.111111 square yard
- **144** square inches
- **0.09290304** square meter
- **929.0304** square centimeters
- **92,903.04** square millimeters

1 *square inch* (in^2) =
- 0.0007716 square yard
- 0.006944 square foot
- 0.00064516 square meter
- **6.4516** square centimeters
- **645.16** square millimeters

1 *square mil* (mil^2) =
- **0.000001** square inch
- **0.00064516** square millimeter

Measure Used for Diameters and Areas of Electric Wires

1 *circular inch* =
- area of 1-inch diameter circle
- $\pi/4$ square inch
- 0.7854 square inch
- 5.067 square centimeters
- 1,000,000 circular mils

1 *circular mil* =
- area of 0.001-inch diameter circle
- $\pi/4$ square mil

1 *square inch* =
- 1.2732 circular inches
- 1,273,239 circular mils

Note: Figures in **Bold** indicate exact conversion values

Table 5. Cubic Measure and Conversion Factors

Metric System

1 *cubic meter* (m^3) =
- **1000** cubic decimeters (liters)
- **1,000,000** cubic centimeters
- 1.30795 cubic yards
- 35.314667 cubic feet
- 61,023.74 cubic inches
- 264.17205 US gallons
- 219.96925 British Imperial gallons

1 *liter (l) or 1 cubic decimeter* (dm^3) =
- **1** liter = volume of 1 kg water at 39.2° F
- **0.001** cubic meter
- **1000** cubic centimeters
- **10** deciliters
- 0.03531466 cubic foot
- 61.023744 cubic inches
- 0.2642 US gallon
- 0.21997 British Imperial gallon
- 1.0566882 US quarts
- 33.814 US fluid ounces

1 *cubic centimeter* (cm^3) =
- **0.001** liter
- **1000** cubic millimeters
- 0.061024 cubic inch

1 *cubic millimeter* = **0.001** cubic centimeters
1 *hectoliter (hl)* = **100** liters
1 *deciliter (dl)* = **10** centiliters
1 *centiliter (cl)* = **10** milliliters

US Customary System

1 *cubic yard* (yd^3) =
- **27** cubic feet
- 201.97403 US gallons
- **46,656** cubic inch
- 0.7646 cubic meter

1 *cubic foot* (ft^3) =
- **1728** cubic inches
- 7.4805 US gallons
- 6.23 British Imperial gallons
- 0.02831685 cubic meter
- 28.31685 liters

1 *cubic inch* (in^3) =
- 0.55411256 US fluid ounce
- **16.387064** cubic centimeters

Shipping Measure

For measuring internal capacity of a vessel:

1 *register ton* = 100 cubic feet

For measurement of cargo:

1 *shipping ton* =

Approximately 40 cubic feet of merchandise is considered a shipping ton, unless that bulk would weigh more than 2000 pounds, in which case the freight charge may be based upon weight

40 *cubic feet* = 32.143 US bushels = 31.16 Imperial bushels

British (Imperial) Liquid and Dry Measure

1 *British Imperial gallon* =
- 0.1605 cubic foot
- 277.42 cubic inches
- 1.2009 US gallon
- **160** Imperial fluid ounces
- **4** Imperial quarts
- **8** Imperial pints
- **4.54609** liters

1 *quart* =
- **2** Imperial pints
- **8** Imperial gills
- **40** Imperial fluid ounces
- 69.354 cubic inches
- **1.1365225** liters

1 *pint* =
- **4** Imperial gills
- **20** Imperial fluid ounces
- 34.678 cubic inches
- **568.26125** milliliters

1 *gill* =
- **5** Imperial fluid ounces
- 8.669 cubic inches
- 142.07 milliliters

US Liquid Measure

1 *US gallon* =
- 0.13368 cubic foot
- **231** cubic inches
- **128** US fluid ounces
- **4** US quarts
- **8** US pints
- 0.8327 British Imperial gallon
- **3.785411784** liters

1 *quart* =
- **2** US pints
- **8** US gills
- **32** US fluid ounces
- **57.75** cubic inches
- 0.9463529 liters

1 *pint* =
- **4** US gills
- **16** US fluid ounces
- **28.875** cubic inches
- 473.176 milliliters

1 *gill* =
- **1/2** cup = **4** US fluid ounces
- **7.21875** cubic inches
- 118.29 milliliters

Note: Figures in **Bold** indicate exact conversion values

Table 5. Cubic Measure and Conversion Factors *(Continued)*

British (Imperial) Liquid and Dry Measure

1 *British Imperial fluid ounce* =
- 1.733871 cubic inch
- 1/160 British Imperial gallon
- 28.41306 milliliters

1 *British Imperial bushel* =
- **8** Imperial gallons = 1.284 cubic feet
- 2219.36 cubic inches

Apothecaries' Fluid Measure

1 *US fluid ounce* =
- **1.8046875** cubic inches
- 1/128 US gallon
- **8** drachms
- 0.02957353 liter
- 29.57353 milliliters

1 *fluid drachm* = **60** minims

US Dry Measure

1 *bushel (US or Winchester struck bushel)* =
- 1.2445 cubic feet
- **2150.42** cubic inches
- a cylinder 18.5 inches dia., 8 inches deep
- a cylinder 47.0 cm dia., 20.3 cm deep

1 *bushel* = **4** pecks = **32** quarts = **64** pints

1 *peck* = **8** quarts = **16** pints

1 *dry quart* = **2** pints =
- **67.200625** cubic inches
- 1.101221 liters

1 *heaped bushel* = 1¼ struck bushel

1 *cubic foot* = 0.8036 struck bushel

Old Liquid Measure

1 *barrel (bbl)* = 31½ gallons

1 *hogshead* = 2 barrels = 63 gallons

1 *pipe* or *butt* = 2 hogsheads = 4 barrels = 126 gallons

1 *tierce* = 42 gallons

1 *puncheon* = 2 tierces = 84 gallons

1 *tun* = 2 pipes = 3 puncheons

Other Cubic Measure

The following are used for wood and masonry:

1 *cord of wood* = 4 × 4 × 8 feet = 128 cubic feet

1 *perch of masonry* =
- 16½ × 1½ × 1 foot = 24¾ cubic feet

Barrel Measure

1 *drum* =
- **55** US gallon
- 7.3524 cubic feet
- 208.19765 liters

1 *petroleum barrel (bo)* =
- **42** US gallons
- 5.614583 cubic feet
- 158.98729 liters

Table 6. Mass and Weight Conversion Factors

Metric System

1 *metric ton (t)* =
- **1000** kilograms
- 2204.6223 pounds
- 0.9842 gross or long ton (of 2240 pounds)
- 0.9072 net or short ton (of 2000 pounds)

1 *kilogram (kg)* =
- **1000** grams = **10** hectograms
- 2.2046 pounds
- 35.274 ounces avoirdupois

1 *hectogram (hg)* = **10** dekagrams

1 *dekagram (dag)* = **10** grams

1 *gram (g)* =
- **10** decigrams
- 0.0022046 pound
- 0.03215 ounce Troy
- 0.03527 ounce avoirdupois
- 15.432 grains

1 *decigram (dg)* = **10** centigrams

1 *centigram (cg)* = **10** milligrams

Avoirdupois or Commercial Weight

1 *gross or long ton* =
- **2240** pounds
- 1.016 metric ton
- 1016 kilograms

1 *net or short ton* = **2000** pounds

1 *pound* =
- **16** ounces
- **7000** grains
- **0.45359237** kilogram
- 453.6 grams

1 *ounce* =
- 1/16 pound
- 16 drachms
- **437.5** grains
- 28.3495 grams
- 0.2780139 newton

1 *grain Avoirdupois* =
- **1** grain apothecaries' weight =
- **1** grain Troy weight
- 0.064799 gram

Note: Figures in **Bold** indicate exact conversion values

Table 6. Mass and Weight Conversion Factors *(Continued)*

Troy Weight

Used for Weighing Gold and Silver

1 *pound Troy* =
12 ounces Troy = 5760 grains
$^{144}/_{175}$ avoirdupois pound

1 *ounce Troy* =
20 pennyweights = 480 grains
31.103 grams

1 *pennyweight* = 24 grains

1 *grain Troy* =
1 grain avoirdupois
1 grain apothecaries' weight
0.0648 gram

1 *carat (used in weighing diamonds)* =
3.086 grains
200 milligrams = $^1/_5$ gram

1 *gold karat* = $^1/_{24}$ proportion pure gold

Apothecaries' Weight

1 *pound* = 12 ounces = 5760 grains

1 *ounce* =
8 drachms = 480 grains
31.103 grams

1 *drachm* = 3 scruples = 60 grains

1 *scruple* = 20 grains

Old Weight Measures

Measures for weight seldom used in the United States:

1 *gross or long ton* = 20 hundred-weights

1 *hundred-weight* = 4 quarters = 112 pounds

1 *quarter* = 28 pounds

1 *stone* = 14 pounds

1 *quintal* = 100 pounds

Table 7. Pressure and Stress Conversion Factors

1 *kilogram per sq. millimeter* (kg_f/mm^2) =
1422.32 pounds per square inch

1 *kilogram per sq. centimeter* (kg_f/cm^2) =
14.223 pounds per square inch

1 *bar* =
1,000,000 dynes per square centimeter
1000 millibars
100 kilopascals
750.06168 torr
1.0197162 kg force per cm^2
14.50377 pounds per square inch
29.529983 inches of mercury at 0° C
10,197.162 mm water at 4° C
33.455256 feet of water at 4° C

1 *millibar* =
100,000 dynes per square centimeter
100 pascal

1 *torr* =
760 millimeters mercury
$^1/_{760}$ atmosphere
133.224 pascal
1.333224 millibar

1 *pound per square inch* =
144 pounds per square foot
0.068 atmosphere
2.042 inches of mercury at 62° F
27.7 inches of water at 62° F
2.31 feet of water at 62° F
0.0703 kilogram per square centimeter
6.894757 kilopascals
6894.757 pascal

1 *atmosphere* =
30 inches of mercury at 62° F
14.7 pounds per square inch
2116.3 pounds per square foot
33.95 feet of water at 62° F

1 *foot of water at 62° F* =
62.355 pounds per square foot
0.433 pound per square inch

1 *inch of mercury at 62° F* =
1.132 foot of water
13.58 inches of water
0.491 pound per square inch

1 *inch of water* =
0.0735559 inch mercury at 0° C
1.8683205 torr
0.5780367 ounce force per square inch
0.0024583 atmosphere

Note: Figures in **Bold** indicate exact conversion values

Table 8. Temperature Conversion Fomulas

To Convert	To	Use Formula
Celsius, t_C	K, t_K	$t_K = t_C + 273.15$
	°F, t_F	$t_F = 1.8\,t_C + 32$
	°R, t_R	$t_R = 9(t_C + 273.15)/5$
Fahrenheit, t_F	K, t_K	$t_K = (t_F + 459.67)/1.8$
	°C, t_C	$t_C = (t_F - 32)/1.8$
	°R, t_R	$t_R = t_F + 459.67$
Kelvin, t_K	°C, t_C	$t_C = t_K - 273.15$
	°F, t_F	$t_F = 1.8\,t_K - 459.67$
	°R, t_R	$t_R = 9/5 \times t_K$
Rankine, t_R	K, t_K	$t_K = 5/9 \times t_R$
	°C, t_C	$t_C = 5/9 \times t_R - 273.15$
	°F, t_F	$t_F = t_R - 459.67$

Kelvin temperatures are expressed by K without use of the degree symbol (°).

Table 9. Energy Conversion Factors

1 *horsepower-hour =*
0.746 kilowatt-hour
1,980,000 foot-pounds
2545 Btu (British thermal units)
2.64 pounds of water evaporated at 212° F
17 pounds of water raised from 62° to 212° F

1 *kilowatt-hour =*
1000 watt-hours
1.34 horsepower-hour
2,655,200 foot-pounds
3,600,000 joules
3415 Btu
3.54 pounds of water evaporated at 212° F
22.8 pounds of water raised from 62° to 212° F

Table 10. Power Conversion Factors

1 *horsepower =*
746 watts
0.746 kilowatt
33,000 foot-pounds/minute
550 foot-pounds/second
2545 Btu/hour
42.4 Btu/minute
0.71 Btu/second
2.64 pounds of water evaporated per hour at 212° F

1 *kilowatt =*
1000 watts
1.34 horsepower
2,654,200 foot-pounds/hour
44,200 foot-pounds/minute
737 foot-pounds/second
3415 Btu/hour
57 Btu/minute
0.95 Btu/second
3.54 pounds of water evaporated per hour at 212° F

1 *watt =*
1 joule/second
0.00134 horsepower
0.001 kilowatt
3.42 Btu/hour
44.22 foot-pounds/minute
0.74 foot-pounds/second
0.0035 pound of water evaporated per hour at 212° F

Table 11. Heat Conversion Factors

1 *Btu (British thermal unit) =*
1052 watt-seconds
778 foot-pounds
0.252 kilogram-calorie
0.000292 kilowatt-hour
0.000393, horsepower-hour
0.00104 pound of water evaporated at 212° F

1 *kilogram calorie =*
3.968 Btu

1 *foot-pound =*
1.36 joules
0.000000377 kilowatt-hour
0.00129 Btu
0.0000005 horsepower-hour

1 *kilogram-meter =*
7.233 foot-pounds

1 *joule =*
1 watt-second
0.00000078 kilowatt-hour
0.00095 Btu
0.74 foot-pound

1 *therm =*
100,000 Btu (US)
29.3 kilowatt-hour
105.5 megajoule

INDEX

A

B

C

D

E

F

G

H

I

J

K

L

M

N

O

P

Q

R

S

T

U

V

W

NOTES

NOTES

www.ingramcontent.com/pod-product-compliance
Lightning Source LLC
Chambersburg PA
CBHW050000260925
33194CB00024B/321

9780831150327